OCEANOGRAPHY AND OCEAN ENGINEERING

W0091597

ECOSYSTEM ASSESSMENT FOR MARINE RESOURCE MANAGEMENT

Maria P. Toscano
EDITOR

Nova Science Publishers, Inc.
New York

Library of Congress Cataloging-in-Publication Data

Ecosystem assessment for marine resource management / editor, Maria P. Toscano.
    p. cm.
  Includes index.
  ISBN 978-1-61470-805-6 (softcover)
  1. Marine ecosystem management. 2. Ecological assessment (Biology) 3. Fishery management. I. Toscano, Maria P. II. United States. National Oceanic and Atmospheric Administration
  QH541.5.S3E268 2011
  333.95'6--dc23
                            2011026364

Published by Nova Science Publishers, Inc. † New York

ECOSYSTEM ASSESSMENT FOR MARINE RESOURCE MANAGEMENT

OCEANOGRAPHY AND OCEAN ENGINEERING

Additional books in this series can be found on Nova's website under the Series tab.

Additional E-books in this series can be found on Nova's website under the E-book tab.

CONTENTS

PREFACE

This book explores the various nationwide commissions which have reviewed and highlighted the importance of incorporating ecosystem principles in ocean and coastal resource management. An ecosystem approach to management (EAM) is one that provides a comprehensive framework for marine, coastal and Great Lakes resource decision making. Integrated ecosystem assessments (IEAs) are a critical science-support element enabling an EAM strategy. An IEA is a formal synthesis and quantitative analysis of information on relevant natural and socioeconomic factors in relation to specified ecosystem management goals. It involves and informs citizens, industry representatives, scientists, resource managers and policy makers through formal processes to contribute to attaining the goals of EAM.

Chapter 1- The reports of the U.S. Oceans Commission, the Pew Oceans Commission, the Ocean Priorities Plan, and other nationwide reviews highlight the importance of incorporating ecosystem principles in ocean and coastal resource management. An ecosystem approach to management (EAM) is one that provides a comprehensive framework for marine, coastal, and Great Lakes resource decision making. Integrated ecosystem assessments (IEAs) are a critical science-support element enabling an EAM strategy. An IEA is a formal synthesis and quantitative analysis of information on relevant natural and socioeconomic factors in relation to specified ecosystem management goals. It involves and informs citizens, industry representatives, scientists, resource managers, and policy makers through formal processes to contribute to attaining the goals of EAM.

Chapter 2- Section 406 of the 2006 Magnuson Stevens Fishery Conservation and Management Reauthorization Act charged NMFS, in consultation with the Fishery Management Councils, to undertake a study on the "state of the science for advancing the concepts and integration of ecosystem considerations in regional fishery management." Section 406 specifies four objectives: 1) form recommendations for scientific data, information, and technology requirements for understanding ecosystem processes and methods for integrating this information from federal, state, and regional sources; 2) form recommendations for processes for incorporating broad stakeholder participation; 3) form recommendations for processes to account for effects of environmental variation on fish stocks and fisheries; and 4) describe existing and developing Council efforts to implement ecosystem approaches, including lessons learned by the Councils.

Chapter 3- MAFAC envisions a future with healthy, sustainable fish populations, a robust fishing and marine offshore aquaculture industry, ample recreational fishing opportunities, numerous, vibrant coastal fishing communities, and a safe and healthy seafood supply for the nation. To achieve this vision, the following recommendations are proposed. (More specific details and rationale for each are found in the Appendices of the report.)

In: Ecosystem Assessment for Marine Resource...   ISBN: 978-1-61470-805-6
Editors: Maria P. Toscano   © 2012 Nova Science Publishers, Inc.

Chapter 1

INTEGRATED ECOSYSTEM ASSESSMENTS

National Oceanic and Atmospheric Administration

EXECUTIVE SUMMARY

The reports of the U.S. Oceans Commission, the Pew Oceans Commission, the Ocean Priorities Plan, and other nationwide reviews highlight the importance of incorporating ecosystem principles in ocean and coastal resource management. An ecosystem approach to management (EAM) is one that provides a comprehensive framework for marine, coastal, and Great Lakes resource decision making. Integrated ecosystem assessments (IEAs) are a critical science-support element enabling an EAM strategy. An IEA is a formal synthesis and quantitative analysis of information on relevant natural and socioeconomic factors in relation to specified ecosystem management goals. It involves and informs citizens, industry representatives, scientists, resource managers, and policy makers through formal processes to contribute to attaining the goals of EAM.

An IEA uses approaches that determine the probability that ecological or socioeconomic properties of systems will move beyond or return to within acceptable limits as defined by management objectives. An IEA must provide an efficient, transparent means of summarizing the status of ecosystem components, screening and prioritizing potential risks, and evaluating alternative management strategies against a backdrop of environmental variability. To this end, IEAs should follow five steps:

1. Scoping. Identify management objectives, articulate the ecosystem to be assessment, identify ecosystem attributes of concerns, and identify stressors relevant to the ecosystem being examined.
2. Indicator development. Researchers develop and test indicators that reflect the ecosystem attributes and stressors specified in the scoping process. Specific indicators are dictated by the problem at hand and must be linked objectively to decision criteria.
3. Risk analysis. This analysis is hierarchical in approach and moves from a comprehensive, but initially qualitative analysis, through a more focused and semiquantitative approach, and finally to a highly focused and fully quantitative approach. The goal of these risk analyses is to fully explore the susceptibility of an indicator to natural or human threats, as well as the ability of the indicator to return to its previous state after being perturbed.
4. Overall ecosystem assessment. Results from the risk analysis for each ecosystem indicator are then integrated in the assessment phase of the IEA. The assessment quantifies the overall status of the ecosystem relative to historical status and prescribed targets.
5. Evaluation. The final phase of the IEA evaluates the potential of different management strategies to influence the status of the ecosystem.

IEAs compel decision makers to squarely confront both the spatial and temporal scales over which ecosystem dynamics, management issues, and societal impacts occur. Scales must be consistent with the ability to recognize and explain the most important drivers and threats to the ecosystem.

There is a clear need to actively integrate diverse physical, biological, and socioeconomic data and think critically about the ways in which decisions affect the ecosystem goods and services that society values. The IEA process we describe accomplishes this task, and provides critical assessment support to the institutional framework upholding societal interests in healthy and productive ecosystems.

ACKNOWLEDGMENTS

The authors thank the members of the Integrated Ecosystem Assessment Priority Task Team for comments on this paper. Discussions with Dave Fluharty, Mary Ruckelshaus, Tony Smith, Chris Harvey, Isaac Kaplan, Jameal

Samhouri, Beth Fulton, John Stein, Chris Costello, Sarah Lester, Steve Gaines, Heather Tallis, and numerous participants of IEA workshops helped sharpen our thinking. We are extremely grateful to Lora Clarke for her hard work in creating this technical memorandum. Phillip Levin also thanks Alberto Contrador for inspiration over the final hurdles.

ABBREVIATIONS AND ACRONYMS

DPSIR	Driver-Pressure-State-Impact-Response
EAM	Ecosystem Approach to Management
IEA	Integrated Ecosystem Assessment
MSE	Management Strategy Evaluation
NOAA	National Oceanic and Atmospheric Administration

BACKGROUND

Reports of the U.S. Oceans Commission, the Pew Oceans Commission, the Ocean Priorities Plan, and other nationwide reviews highlight the importance of incorporating ecosystem principles in ocean and coastal resource management. Specific to the National Oceanic and Atmospheric Administration (NOAA), a critical objective is to "protect, restore, and manage the use of coastal, ocean, and Great Lakes resources through an Ecosystem Approach to Management (EAM)" (NOAA 2005). An ecosystem approach to management is one that provides a comprehensive framework for marine, coastal, and Great Lakes resource decision making. In contrast to individual species or single issue management, an EAM considers a wider range of ecological, environmental, and human factors bearing on diverse societal objectives regarding resource use and protection.

What Is an Ecosystem?

As defined by NOAA:
"An ecosystem is a geographically specified system of organisms (including humans), the environment, and the processes that control its dynamics" (Murawski and Matlock 2006). NOAA further defines the environment as "the biological, chemical, physical, and social

conditions that surround organisms. When appropriate, the term environment should be qualified as biological, chemical, and/or social" (Murawski and Matlock 2006).

What is an Integrated Ecosystem Assessment?

An Integrated Ecosystem Assessment (IEA) is a critical science-support element enabling an EAM strategy. An IEA is a formal synthesis and quantitative analysis of information on relevant natural and socioeconomic factors relative to specified ecosystem management goals. It involves and informs citizens, industry representatives, scientists, resource managers, and policy makers through formal processes to contribute to attaining the goals of EAM. In this technical memorandum, we outline a stepwise approach that will guide the science of IEAs.

IEAs begin with an identification of critical management and policy questions to define the scope of information and analyses necessary to inform management. IEAs use quantitative analyses and ecosystem modeling to integrate a range of social, economic, and natural science data and information to assess the condition of the ecosystem relative to the identified scope. IEAs also identify potential management options and these are evaluated against EAM goals. IEAs are peer-reviewed and communicated to stakeholders, resource managers, and policy makers. IEAs differ from other assessments such as Environmental Impact Statements in that they explicitly consider all components of the ecosystem and address the broad goals of EAM.

An IEA consists of the following components:

- Identification of key issues of concerns and stressors that management and policy should address
- Assessment of status, indicators, and trends of the ecosystem condition relative to established management targets or thresholds
- Assessment of the environmental, social, and economic causes and consequences of these trends
- Forecast of ecosystem condition under a range of policy or management actions or both
- Periodic reevaluation of management effectiveness in the context of emerging ecosystem issues
- Identification of crucial knowledge and data gaps that will guide future research and data acquisition efforts

Why IEAs?

A key goal of IEAs is to move toward clear, well-defined ecosystem objectives built on a science strategy that fuses ecosystem components into a single, dynamic fabric in which human and natural factors are intertwined. Periodic assessment of biological, chemical, physical, and socioeconomic attributes of ecosystems allows for coordinated evaluations of national marine, coastal, and Great Lakes ecosystems to promote their sustainability under a variety of human uses and environmental stresses. Moreover, IEAs involve and inform a wide variety of stakeholders and agencies that rely on science support. IEAs integrate knowledge and data collected by NOAA and regional entities including other federal agencies, states, nongovernmental organizations, and academic institutions. IEAs also identify critical knowledge and data gaps, which if filled will reduce uncertainty and improve our ability to fully employ ecosystem approaches to management.

The Importance of Scale

IEAs must explicitly consider both spatial extent and time domains over which ecosystem dynamics and management issues occur. Scales must be consistent with the ability to recognize and explain the most important drivers and threats to the ecosystem. Ecosystems typically do not have sharp boundaries; rather, one ecosystem blends into another. As a consequence, ecosystem boundaries are human constructs, and a first step in any IEA endeavor must be to identify the spatial scale of the problem under consideration. The spatial scale of an IEA is a function of the ecology, geology, and oceanography of a region, as well as the scale of management issues and governance structures. For example, while an IEA may focus on a small embayment, consideration of large-scale issues such as climatic variability as well as linkages to adjacent ecosystems are important. IEAs should address the linkage of terrestrial, coastal, and oceanic environments as part of or affecting the ecosystem. Additionally, IEAs must be cognizant of appropriate temporal scales. In particular, IEAs require attention to the temporal baseline against which current status is compared. For example, different conclusions may be drawn when the comparing current ecosystem conditions to those of 25 years versus 75 years ago.

APPLYING THE INTEGRATED ECOSYSTEM ASSESSMENT CONCEPT

An IEA uses approaches that determine the probability that ecological or socioeconomic properties of systems will move beyond or return to within acceptable limits as defined by management objectives. An IEA must provide an efficient, transparent means of summarizing the status of ecosystem components, screening and prioritizing potential risks, and evaluating alternative management strategies against a backdrop of environmental (e.g., climatic, oceanographic, seasonal, real-time weather) variability. An IEA provides a means of evaluating trade-offs in management strategies among potentially competing ecosystem use sectors.

A Five Step Process for an IEA

The five step IEA process discussed here is comprised of scoping, indicator identification and testing, risk analysis, risk analysis integration into the assessment process, and strategies evaluation. The process is conceptually portrayed in Figure 1.

Step 1. A scoping process initiates the IEA. Scoping begins with a review of existing documents and information and concludes with stakeholder, resource manager, and policy maker involvement to identify the management objectives, articulate the ecosystem to be assessed, identify ecosystem attributes of concern, and identify stressors relevant to the ecosystem being examined. While general EAM goals may be broad, a key component of an IEA is to move from broad goals to specific ecosystem objectives that management and policy makers need to consider.

Step 2. Following the scoping process, researchers develop and test indicators that reflect the ecosystem attributes and stressors specified in the scoping process. Specific indicators are dictated by the problem at hand and must be linked objectively to decision criteria. In some cases, this simply means following the abundance of a single species (e.g., in the case of an endangered species) or suites of species (e.g., coral reefs, harmful algal blooms). In other instances, the indicator may be a proxy for an ecosystem "attribute" indicated in Step 1. For example, resiliency to perturbation might be an attribute and species diversity might be an "indicator" of resiliency. For many problems, suites of indicators that span a wide range of processes (with different associated rates), biological groups, and indicator types (e.g., "early warning," "integrated system state") will be necessary. Importantly, this step allows us to identify indicators that should be monitored even when current monitoring efforts are insufficient.

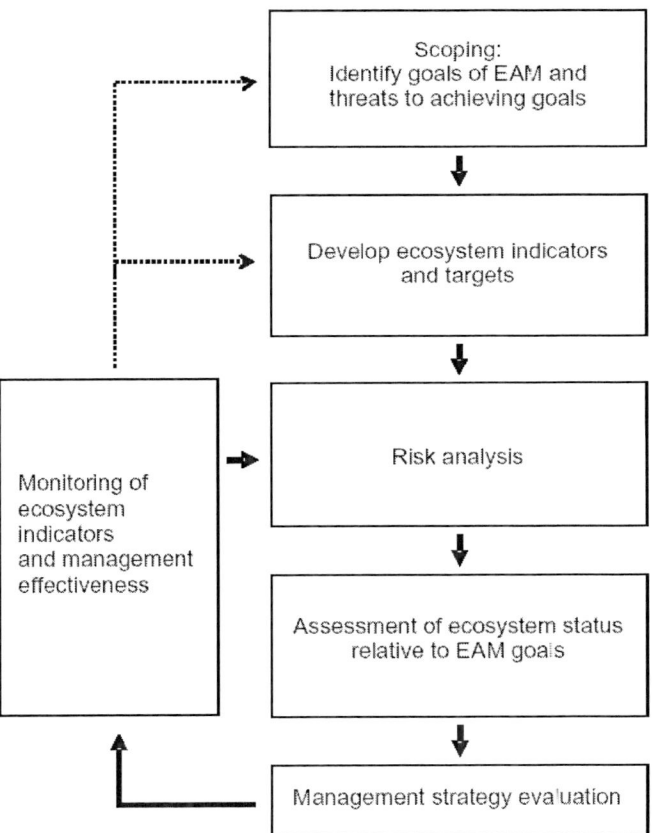

Figure 1. A five step process for an IEA. The solid line from the monitoring box to the risk analysis box indicates that analyses will be updated as more data become available. The dotted lines to the scoping box and the indicators and targets box indicate that these steps may need revisiting as more data are collected.

Step 3. Once indicators are chosen, an analysis is performed that evaluates the risk to the indicators posed by human activities and natural processes. This analysis is hierarchical in approach and moves from a comprehensive, but initially qualitative analysis, through a more focused and semiquantitative approach, and finally to a highly focused and fully quantitative approach. This step initially screens out many potential risks, so that more intensive and quantitative analyses are limited to a subset of ecosystem indicators and human or natural threats. The goal of these risk analyses is to fully explore the susceptibility of an indicator to natural or human threats as well as the ability of the indicator to return to its previous state after being perturbed. Another goal of the risk analyses is to explain, if the indicator has settled at a new value, whether the new value is due to natural variability in the system. A full discussion of ecological risk analysis as it pertains to marine ecosystems can be found in Hobday et al. (2006).

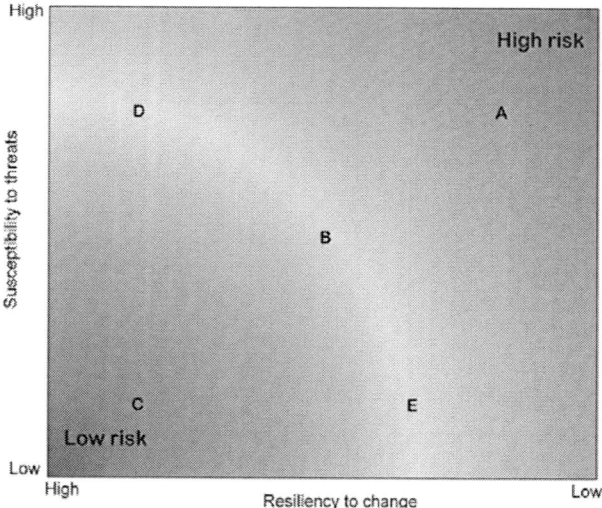

Figure 2. A visualization of the risk status of theoretical indicators. Indicator A has a low resiliency (ability to persist in the face of change) and high susceptibility to natural or human disturbance; thus it has a high risk. Indicator C shows an indicator with low susceptibility and high resilience and thus has low risk. Indicators B, D, and E have difference combinations of susceptibility and resilience yielding a moderate level of risk.

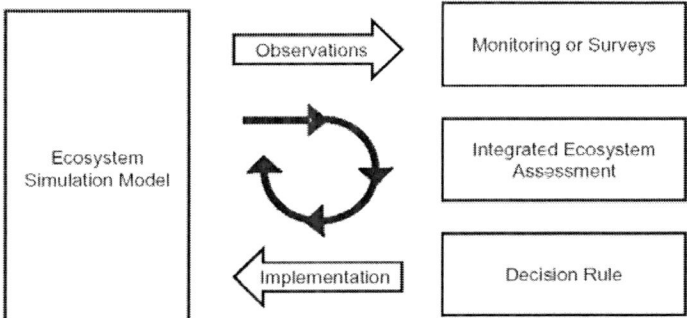

Figure 3. A schematic of the MSE. An ecosystem model is used to simulate the ecosystem. The ecosystem is then "sampled," an IEA is performed, and a management strategy is implemented. The cycle is then repeated, and ultimately the potential outcomes of a range of management strategies can be estimated.

The likelihood of an ecosystem changing can be viewed as the relationship of the susceptibility of a particular indicator to impact versus the resiliency of the indicator (Figure 2). An indicator is likely to change when susceptibility to impact is high and resiliency is low, while an indicator is not likely to change when susceptibility to impact is low and resiliency is high. IEAs will also include a social and economic overlay to the ecological risk assessments to capture impacts to individuals and communities.

Step 4. Results from the risk analysis for each ecosystem indicator are then integrated in the assessment phase of the IEA. The assessment quantifies the status of the ecosystem relative to historical status and prescribed targets. Thus the risk analysis rigorously quantifies the status of individual ecosystem indicators, while the full assessment considers the state of all indicators simultaneously.

Step 5. The next phase of the IEA uses ecosystem modeling frameworks (e.g., the Atlantis ecosystem model, Brand et al. 2007) to evaluate the potential of different management strategies to influence the status of natural and human system indicators. To accomplish this, a formal Management Strategy Evaluation (MSE) is employed (Figure 3). In MSE, a simulation model is used to generate "true" ecosystem dynamics. Data are sampled from the model to simulate research surveys, then these data are passed to risk analysis and assessment models. These assessment models estimate the predicted status of individual indicators and the ecosystem as a whole. Based on this assessment of the simulated ecosystem, a management decision is simulated. Human response to this simulated decision is modeled and potentially influences the simulated ecosystem state. By repeating this cycle, we can simulate the full management cycle. This allows us to test the utility of modifying indicators and threshold levels, assessments, monitoring plans, management strategies, and decision rules. MSE in the context of an IEA can thus serve as a filter to identify which policies and methods meet stated management objectives (e.g., Butterworth and Punt 1999).

IEA Products

IEAs are peer-reviewed and communicated to stakeholders, resource managers, and policy makers. IEAs may be communicated in the form of a static MSE framework or as Web-based dynamic documents that are updated as new data become available. The frequency with which IEAs should be revised and updated cannot be fully prescribed. As new information arises or management changes occur, risks can be reevaluated and documented as before. IEA products may also serve as a tool to educate a variety of stakeholders.

Further Reading on IEAs

The concept of IEAs is well established and a number of examples are relevant in both domestic and international settings. Appendix B provides a selected, annotated bibliography of existing IEA documents as well as Web-based resources describing the concept in more detail.

REFERENCES

[1]   Brand, E. J., Kaplan, I. C., Harvey, C. J., Levin, P. S., Fulton, E. A., Hermann, A. J. & Field, J. C. (2007). *A spatially explicit ecosystem model of the California Current's food web and oceanography.* U.S. Dept. Commer., NOAA Tech. Memo. NMFS-NWFSC-84.

[2]   Butterworth, D. S. & Punt, A. E. (1999). Experiences in the evaluation and implementation of management procedures. *ICES J. Mar. Sci, 56*, 985–998.

[3]   Hobday, A. J., Smith, A., Webb, H., Daley, R., Wayte, S., Bulman, C., Dowdney, J., Williams, A., Sporcic, M., Dambacher, J., Fuller, M. & Walker, T. (2006). *Ecological risk assessment for the effects of fishing: Methodology.* Rep. R04/1072 for the Australian Fisheries Management Authority, Canberra.

[4]   Murawski, S. A. & Matlock, G. C. (eds.). (2006). *Ecosystem science capabilities required to support NOAA's mission in the year 2020.* U.S. Dept. Commer., NOAA Tech. Memo. NMFS-F/SPO-74.

[5]   NOAA (National Oceanic and Atmospheric Administration). (2005). New priorities for the 21st century—NOAA's strategic plan: Updated for FY 2006-FY 2011. U.S. Dept. Commer., NOAA, Silver Spring, MD. Online at http://www.ppi.noaa.gov/PPI_Capabilities/Documents/ Strategic_Plans/ FY06-11_NOAA_Strategic_Plan.pdf [accessed 27 June 2008].

APPENDIX A. THE DPSIR FRAMEWORK

The strategy described in this technical memorandum can be cast in the context of a Driver-Pressure-State-Impact-Response (DPSIR) framework for classification of indicators. The DPSIR approach has been broadly applied in environmental assessments of terrestrial and aquatic ecosystems.

Drivers are factors that result in pressures that in turn cause changes in the system. For the purposes of an Integrated Ecosystem Assessment (IEA), both natural and anthropogenic forcing factors are considered; an example of the former is climate variability while the latter include factors such as human population size in the coastal zone, associated coastal development, demand for seafood, etc. In principle, human driving forces can be assessed and controlled. Natural environmental changes cannot be controlled but must be accounted for in management.

Pressures include factors such as coastal pollution, habitat loss and degradation, and fishing effort that can be mapped to specific drivers. For example, coastal development results in increased coastal armoring and the loss of associated intertidal habitat.

State variables are indicators of the condition of the ecosystem (including physical, chemical, and biotic factors).

Impacts comprise measures of the effect of change in these state variables such as loss of biodiversity, declines in productivity and yield, etc. Impacts are measured with respect to management objectives and the risks associated with exceeding or returning to below these targets and limits.

Responses are the actions (regulatory and otherwise) that are taken in response to predicted impacts. Forcing factors under human control trigger management responses when target values are not met as indicated by risk assessments. Natural drivers may require adaptational response to minimize risk. For example, changes in climate conditions that in turn affect the basic productivity characteristics of a system may require changes in ecosystem reference points that reflect the shifting environmental states.

Table A-1. Examples of DPSIR components of interest for IEAs

Components	Anthropogenic	Natural
Drivers	Human population size in the coastal zone Per capita seafood demand Water-dependent international trade Coastal development	Temperature Precipitation Winds Ice cover Hydrodynamics
Pressures	Fishing effort Habitat loss and degradation Pollution transport and fate Marine transportation Effluent discharges Oil and hazardous material spills Pathogens Land use patterns	Extent of thermal habitat Nutrient regeneration Current speed and direction Habitat change Species range shifts
States	Commercial fishery landings Recreational fishery landings Aquaculture and fish farming production Water quality and quantity	Chlorophyll concentration Zooplankton biomass Benthic biomass Shellfish biomass Fish biomass Harmful algal blooms Pathogens Aquatic mammal abundance
Impacts	Fishery yield Aquaculture production Recreational income Nonindigenous species Human health risks Employment Loans at risk Commercial cash value	Biodiversity Changes in ecosystem function
Responses	Alter fishing mortality Alter stormwater regulations Require watershed buffers Restore habitat Contaminant mitigation	

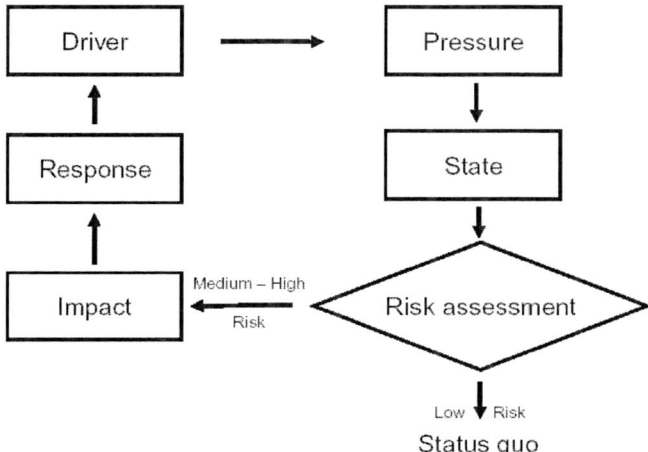

Figure A-1. The DPSIR framework for classification of ecosystem indicators. The process begins with the driver box.

The different classes of indicators identified within the DPSIR framework can be mapped to the needs of the management strategy evaluation described above and identified with respect to their roles in model formulation, parameterization, and validation. Table A-1 provides examples of DPSIR components. Figure A-1 shows the conceptual framework.

APPENDIX B. BIBLIOGRAPHY

The following is a bibliography, annotated for many documents, of integrated ecosystem assessment concepts, methods, evaluations, and implementation examples.

Conceptual Framework Documents

DFO Canada (Dept. Fisheries and Oceans Canada). Canada's oceans strategy: Policy and operational framework for integrated management of estuarine, coastal, and marine environments in Canada. 2002. Dept. Fisheries and Oceans Canada, Oceans Directorate, Ottawa, Ontario. Online at http://www.dfo-mpo.gc.ca/ oceans-habitat/ oceans/ri-rs/index_e.asp [accessed 12 June 2008].

The Canadian Oceans Act calls on the Minister of Fisheries and Oceans to lead and facilitate the development of a national oceans strategy that will guide the management of Canada's estuarine, coastal, and marine ecosystems. This strategy provides the overall strategic framework for Canada's oceans-related programs and policies, based on the principles of sustainable development, integrated management, and precautionary approach. The central governance mechanism of the strategy is applying these principles through the development and implementation of integrated management plans. This document is intended to foster discussion about integrated management approaches by setting out policy in the legislative context, along with concepts and principles. The document also proposes an operational framework with governance, management by areas, design for management bodies, and the type of planning processes that could be involved.

Link, J. S. 2005. Translating ecosystem indicators into decision criteria. ICES J. Mar. Sci. 62:569–576.

Defining and attaining suitable management goals probably represent the most difficult part of ecosystem-based fisheries management. To achieve those goals, we ultimately need to define ecosystem overfishing in a way that is analogous to the concept used in single-species management. Ecosystem-based control rules can then be formulated when various ecosystem indicators are evaluated with respect to fishing-induced changes. However, these multiattribute control rules will be less straightforward than those applied typically in single-species management, and may represent gradient rather than binary decision criteria. Some ecosystem-based decision criteria are suggested, based on indicators empirically derived from the Georges Bank, Gulf of Maine ecosystem. Further development in the translation of ecosystem indicators into decision criteria is one of the major areas for progress in fisheries science and management.

Rice, J. 2003. Environmental health indicators. Ocean Coast. Manag. 46:235–259.

Sherman, K. 1991. The large marine ecosystem concept: Research and management strategy for living marine resources. Ecol. Appl. 1(4):349–360.

Methods and Tools

DFO Canada (Dept. Fisheries and Oceans Canada). 2006. Identification of ecologically significant species and community properties. DFO Canadian Science Advisory Secretariat, Science Advisory Rep. 2006/041. Online at http://www.dfo-mpo.gc.ca/csas/Csas/status/ 2004/ESR2004_006_E.pdf [accessed 12 June 2008].

As with the criteria described above for ecologically and biologically significant areas, consistent criteria and guidance for their application are needed also for the identification of species and community properties for which protection should be enhanced, while allowing sustainable activities to be pursued in the ecosystem. This report contains the results of a national workshop held in 2006 to develop a priori criteria to assess species and community properties that are "particularly important" or "significant" with regard to maintaining ecosystem structure and function. Assessments using these criteria as a tool to rank species and community properties by their ecological significance are an important step in developing ecosystem objectives for integrated management.

For a recent geographic application of these criteria please see: Dept. Fisheries and Oceans Canada. 2006. Proceedings of the zonal workshop on the identification of ecologically and biologically significant areas within the Gulf of St. Lawrence and estuary. DFO Canadian Science Advisory Secretariat Proceed. Ser. 2006/011. Online at http://www.dfo-mpo.gc.ca/csas/Csas/Proceedings/2006/ PRO2006_011_B.pdf [accessed 12 June 2008].

Edmonds, J. A., and N. J. Rosenberg. 2005. Climate change impacts for the conterminous USA: An integrated assessment. Clim. Change 69:1–162.

This special issue of the journal Climatic Change describes an effort to improve methodology for integrated assessment of impacts and consequences of climatic change. The methodology developed involves construction of climate change scenarios that are used to drive individual sectoral models for simulating impacts on crop production, irrigation demand, water supply, and change in productivity and geography of unmanaged ecosystems. Economic impacts of the changes predicted by integrating the results of the several sectoral simulations models are calculated through an agricultural land use model. While these analyses were conducted for the conterminous United States, their global implications are also considered, as is the need for further improvements in integrated assessment methodology. The final chapter summarizes highlights of the first seven sector-specific chapters that constitute this special issue. These projects were supported by the National Science Foundation through the Methods and Models in Integrated Assessment Program and in some cases also by the U.S. Department of Energy Integrated Assessment Program, Biological and Environmental Research.

Farber, S., R. Costanza, D. L. Childers, J. Erickson, K. Gross, M. Grove, C. S. Hopkinson, J. Kahn, S. Pincetl, A. Troy, P. Warren, and M. Wilson. 2006. Linking ecology and economics for ecosystem management. Bioscience 56:121–133.

ICES (International Council for the Exploration of the Sea). 2004. Supporting European marine integrated ecosystem assessments: Specific support actions. Copenhagen, Denmark. Online at: http://www.ices.dk/globec/regns/SEMIEA.pdf [accessed 12 June 2008].

Link, J. S., C. A. Griswold, E. T. Methratta, and J. Gunnard (eds.). 2006. Documentation for the Energy Modeling and Analysis eXercise (EMAX). Northeast Fisheries Science Center. Ref. Doc. 06-15. Online at http://www.nefsc.noaa.gov/nefsc/publications/series/crdlist.htm [accessed 12 June 2008].

Peirce, M. 1998. Computer-based models in integrated environmental assessment. Tech. Rep. 14, prepared for the European Environment Agency, Copenhagen, Denmark. Online at http://reports.eea.europa.eu/TEC14/en [accessed 12 June 2008].

Shanmuganathan, S., P. Sallis, and J. Buckeridge. 2006. Self-organizing map methods in integrated modeling of environmental and economic systems. Environ. Model. Softw. 21:1247–1256.

Evaluations of Integrated Assessment Products and Processes

Costanza, R., and S. E. Jorgensen (eds.). 2002. Understanding and solving environmental problems in the 21st century: Toward a new, integrated hard problem science. Elsevier Science. Online (description) at http://www.elsevier.com/wps/find/book

description.cws_home/623393/description#description. [accessed 12 June 2008].

DFO Canada (Dept. Fisheries and Oceans Canada). 2005. Guidelines on evaluating ecosystem overview and assessments: Necessary documentation. DFO Canadian Science Advisory Secretariat Rep. 2005/026. Online at http://www.dfo-mpo.gc.ca/csas/Csas/ status/2005/SAR-AS2005_026_E.pdf [accessed 12 June 2008].

The integrated management of human activities on the sea under Canada's Oceans Act calls for implementation strategies based on an ecosystem approach. In planning many of the activities necessary for integrated management, such as setting ecosystem objectives, identifying areas requiring enhanced protection, and developing regulatory approaches to various activities, it is necessary to have a reasonable understanding of the ecosystem being managed. The Department of Fisheries and Oceans has adopted an approach of preparing two types of documents—ecosystem overview reports and ecosystem assessments—to provide a common factual basis for dialogue among the parties in integrated planning and management. Initial ecosystem overview reports and partial integrated ecosystem assessments were prepared for two ecosystems for which integrated management approaches are currently being developed: the Eastern Scotian Shelf and Gulf of St. Lawrence systems. The overview and assessment documents for the two systems were prepared in different ways, allowing the Department of Fisheries and Oceans to report here on insights gained from a review held in 2005 on the desirable contents to be included in both types of documents.

NAPAP (National Acid Precipitation Assessment Program). 1991. Report from the Oversight Review Board. The experience and legacy of NAPAP. National Science and Technology Council, Washington, DC.

Parsons, E. 1995. Integrated assessment and environmental policy making: In pursuit of usefulness. Energ. Policy 23:463–475.

Toth, F. L. 2001. Participatory integrated assessment methods—An assessment of their usefulness to the European Environmental Agency. Tech. Rep. 64, prepared for the European Environment Agency, Copenhagen, Denmark. Online at http://reports .eea.europa.eu/Technical_report_no_64/en [accessed 12 June 2008].

Integrated Assessment Implementation

National Examples

CENR (Committee on Environmental and Natural Resources). 2000. National assessment of harmful algal blooms in U.S. waters. National Science and Technology Council, Washington, DC. Online at http://www.cop.noaa.gov/pubs/habhrca/Nat_Assess   _HABs.pdf [accessed 12 June 2008].

CENR (Committee on Environmental and Natural Resources). 2003. An assessment of coastal hypoxia and eutrophication in U.S. waters. National Science and Technology Council, Washington, D.C. Online at http://coastalscience.noaa.gov/documents/coastalhypoxia   .pdf [accessed 12 June 2008].

Turgeon, D. D., R. G. Asch, B. D. Causey, R. E. Dodge, W. Jaap, K. Banks, J. Delaney, B. D. Keller, R. Speiler, C. A. Matos, J. R. Garcia, E. Diaz, D. Catanzaro, C. S. Rogers, Z. Hillis-Starr, R. Nemeth, M. Taylor, G. P. Schmahl, M. W. Miller, D. A.Gulko, J. E. Maragos, A. M. Friedlander, C. L. Hunter, R. S. Brainard, P. Craig, R. H. Richond, G. Davis, J. Starmer, M. Trianni, P. Houk, C. E. Birkeland, A. Edward, Y. Golbuu, J. Gutierrez, N. Idechong, G. Paulay, A. Tafileichig, and N. Vander Velde. 2002. The state of coral reef ecosystems of the United States and Pacific freely associated states: 2002. National Oceanic and Atmospheric Administration, National Ocean Service, National Centers for Coastal Ocean Science, Silver Spring, MD. Online at http://coastalscience.noaa.gov/ documents/ status_coralreef.pdf [accessed 12 June 2008].

Waddell, J. E. (ed.). 2005. The state of coral reef ecosystems of the United States and Pacific freely associated states: 2005. NOAA/NCCOS Center for Coastal Monitoring and Assessment's Biogeography Team, Silver Spring, MD. NOAA Tech. Memo. NOS NCCOS 11. Online at http://ccma.nos.noaa.gov/ecosystems/ coralreef/coral_report _2005/ [accessed 12 June 2008].

Regional Examples

Boldt, J. (ed.). 2007. Stock assessment and fishery evaluations, Appendix C, Ecosystem considerations for 2008. Alaska Fisheries Science Center, Seattle, WA. Online at http://access.afsc.noaa.gov/ reem/ EcoWeb/content/pdf/AppendixC.pdf [accessed 12 June 2008].

Brown, B. S., W. Munns Jr., and J. F. Paul. 2002. An approach to integrated ecological assessment of resource condition: The Mid-Atlantic estuaries as a case study. J. Environ. Manag. 66:411–427.

CENR (Committee on Environment and Natural Resources). 2000. Integrated assessment of hypoxia in the Northern Gulf of Mexico. National Science and Technology Council, Washington, DC. Online at http://www.nos.noaa.gov/products/pubs_hypox.html#fia [accessed 12 June 2008].

Hare, J. A., and P. E. Whitfield. 2003. An integrated assessment of the introduction of lionfish (*Pterois volitans/P. miles* complex) to the western Atlantic Ocean. NOAA Tech. Memo. NOS NCCOS 2. Online at http://coastalscience.noaa.gov/documents/ lionfish_ia.pdf [accessed 12 June 2008].

Link, J., and J. Brodziak, (eds.). 2002. Report on the status of the northeast U.S. continental shelf ecosystem. Northeast Fisheries Science Center Ref. Doc. 02-11. Online at http://www.nefsc.noaa.gov/ nefsc/publications/crd/crd0211/ [accessed 12 June 2008].

NCCOS (NOAA National Centers for Coastal Ocean Science). 2005. A biogeographic assessment of the Channel Islands National Marine Sanctuary: A review of boundary expansion concepts for NOAA's National Marine Sanctuary Program. NOAA Tech. Memo. NOS NCCOS 21. Online (DVD format) at http://ccmaserver.nos .noaa.gov/products/biogeography/cinms/order.html [accessed 12 June 2008].

NCCOS (NOAA National Centers for Coastal Ocean Science). 2006. An ecological characterization of the Stellwagen Bank National Marine Sanctuary region: Oceanographic, biogeographic, and contaminants assessment. Prepared by NCCOS's Biogeography Team in cooperation with the National Marine Sanctuary Program. NOAA Tech. Memo. NOS NCCOS 45. Online at http://www.ccma.nos. noaa.gov/products/biogeography/stellwagen/welcome.html [accessed 12 June 2008].

U.S. EPA (Environmental Protection Agency). 2005. National coastal condition report II: Chapter 9—Health of Galveston Bay for human use. EPA-620/R-03/002. EPA Office of Research and Development and Office of Water, Washington, DC. Online at *http://www.epa.gov* /owow/oceans/nccr/2005/Chapter9_GalvestonBay.pdf [accessed 12 June 2008].

International Examples

DFO Canada (Dept. Fisheries and Oceans Canada). 2003. State of the eastern Scotian Shelf ecosystem. DFO Canadian Science Advisory Secretariat. Ecosystem Status Rep. 2003/004. Online at http://www. dfo-mpo.gc.ca/csas/Csas/status/2003/ESR2003_004_e.pdf [accessed 12 June 2008].

European Environment Agency. 2005. The European environment—State and outlook 2005. State of Environment Rep. No. 1/2005. European Environment Agency, Copenhagen, Denmark. Online at http: //reports.eea.europa.eu/state_of_environment_report_2005_1/en [accessed 12 June 2008].

European Environment Agency. 2006. The changing faces of Europe's coastal areas. EEA Rep. No. 6/2006. European Environment Agency, Copenhagen, Denmark. Online at http://reports.eea.europa.eu/ eea_report_2006_6/en [accessed 12 June 2008].

ICES (International Council for the Exploration of the Sea). 2006. Report of the Regional Ecosystem Study Group of the North Sea (REGNS), 15-19 May 2006. ICES Headquarters, Copenhagen, Denmark. ICES CM 2006/RMC:06. Online at http://www.ices.dk/ reports/RMC /2006/REGNS/regns06.pdf [accessed 12 June 2008].

This report summarizes the results of a meeting of the study group held in May 2006 to evaluate and prepare plans for finalization of an integrated assessment of the North Sea ecosystem, an activity this group initiated in 2003. The assessment, based on the compilation and analyses of a comprehensive integrated data set, has provided some valuable insights into the significance of the relationships between different human pressures (e.g., nutrient inputs and fisheries) and state changes (e.g., plankton, fish, and seabirds) at different spatial scales and the time scales over which changes take place. For example, plankton community data in relation to the physical and chemical oceanography reveals both gradients of response to the major riverine inputs of nutrients into the North Sea and sources of nutrients from the Atlantic. In addition, an assessment of all variables reveals two relatively stable states in the North Sea, one pre-1983 and the other post-1997. The intervening years are dominated by high ecosystem variability, which represents a transition from one state to another and in part explains the number of studies which highlight different years for regime shifts. The sensitivity of such analysis to changes in temporal and spatial scales is explored, as is the dependency on the number and type of ecosystem variables. By better understanding the relationship between the causes of change at different scales in time and space, it should be possible to set more realistic targets for the management of human pressures.

Recent NOAA Technical Memorandums published by the Northwest Fisheries Science Center

NOAA Technical Memorandum NMFS-NWFSC-

91 Wainwright, T.C., M.W. Chilcote, P.W. Lawson, T.E. Nickelson, C.W. Huntington, J.S. Mills, K.M.S. Moore, G.H. Reeves, H.A. Stout, and L.A. Weitkamp. 2008. Biological recovery criteria for the Oregon Coast coho salmon evolutionarily significant unit. U.S. Dept. Commer., NOAA Tech. Memo. NMFSNWFSC-91, 199 p. NTIS number pending.

90 Ward, L., P. Crain, B. Freymond, M. McHenry, D. Morrill, G. Pess, R. Peters, J.A. Shaffer, B. Winter, and B. Wunderlich. 2008. Elwha River Fish Restoration Plan–Developed pursuant to the Elwha River Ecosystem and Fisheries Restoration Act, Public Law 102-495. U.S. Dept. Commer., NOAA Tech. Memo. NMFSNWFSC-90, 168 p. NTIS number pending.

89 Holt, M. 2008. Sound exposure and Southern Resident killer whales *(Orcinus orca):* A review of current knowledge and data gaps. U.S. Dept. Commer., NOAA Tech. Memo. NMFS-NWFSC-89, 59 p. NTIS number pending.

88 Olson, O.P., L. Johnson, G. Ylitalo, C. Rice, J. Cordell, T. Collier, and J. Steger. 2008. Fish habitat use and chemical contaminant exposure at restoration sites in Commencement Bay, Washington. U.S. Dept. Commer., NOAA Tech. Memo. NMFS-NWFSC-88, 117 p. NTIS number pending.

87 Keller, A.A., B.H. Horness, V.H. Simon, V.J. Tuttle, J.R. Wallace, E.L. Fruh, K.L. Bosley, D.J. Kamikawa, and J.C. Buchanan. 2007. The 2004 U.S. West Coast bottom trawl survey of groundfish resources off Washington, Oregon, and California: Estimates of distribution, abundance, and length composition. U.S. Dept. Commer.,

NOAA Tech. Memo. NMFS-NWFSC-87, 134 p. NTIS number pending.

86 Keller, A.A., V.H. Simon, B.H. Horness, J.R. Wallace, V.J. Tuttle, E.L. Fruh, K.L. Bosley, D.J. Kamikawa, and J.C. Buchanan. 2007. The 2003 U.S. West Coast bottom trawl survey of groundfish resources off Washington, Oregon, and California: Estimates of distribution, abundance, and length composition. U.S. Dept. Commer., NOAA Tech. Memo. NMFS-NWFSC-86, 130 p. NTIS number pending.

85 Norman, K., J. Sepez, H. Lazrus, N. Milne, C. Package, S. Russell, K. Grant, R.P. Lewis, J. Primo, E. Springer, M. Styles, B. Tilt, and I. Vaccaro. 2007. Community profiles for West Coast and North Pacific fisheries–Washington, Oregon, California, and other U.S. states. U.S. Dept. Commer., NOAA Tech. Memo. NMFS-NWFSC-85, 602 p. NTIS number pending.

In: Ecosystem Assessment for Marine Resource...   ISBN: 978-1-61470-805-6
Editors: Maria P. Toscano   © 2012 Nova Science Publishers, Inc.

Chapter 2

THE STATE OF SCIENCE TO SUPPORT AN ECOSYSTEM APPROACH TO REGIONAL FISHERY MANAGEMENT

National Oceanic and Atmospheric Administration

ACKNOWLEDGMENTS

The steering committee for this workshop and report to Congress consisted of Jon Brodziak, Elizabeth Clarke, Ned Cyr, Patricia Livingston, Alec MacCal l, Thomas Noji, and Roger Zimmerman. Rosemary Kosaka, NMFS Office of Science and Technology, provided background material for the fishery management council process. David Detlor and Lora Clarke, NMFS Office of Science and Technology, assisted with the editing of this report.

EXECUTIVE SUMMARY

Section 406 of the 2006 Magnuson Stevens Fishery Conservation and Management Reauthorization Act charged NMFS, in consultation with the Fishery Management Councils, to undertake a study on the "state of the science for advancing the concepts and integration of ecosystem considerations in regional fishery management." Section 406 specifies four objectives: 1) form recommendations for scientific data, information, and technology requirements for understanding ecosystem processes and methods for integrating this information from federal, state, and regional sources; 2) form recommendations for processes for incorporating broad stakeholder participation; 3) form recommendations for processes to account for effects of environmental variation on fish stocks and fisheries; and 4) describe existing and developing Council efforts to implement ecosystem approaches, including lessons learned by the Councils.

Regarding objective 1, the most important action should be to maintain and expand current fishery- dependent and fishery-independent surveys. These surveys provide the critical information on exploited and unexploited species required to support stock assessments, as well as long-term data on ecosystem status and trends. Most current surveys do not provide sufficient information to effectively manage all stocks, and there is a particular need to increase their spatial and temporal coverage. Additional time-series data on benthic environments are also needed to improve understanding of the relationship between habitat, benthic organisms, and fish species. Increased socioeconomic surveys are needed to help us understand and predict the behavior of harvesters, an important component of the ecosystem, with regard to different management options. Although we need improved ecological models to better understand dynamic ecosystem processes, in many cases modelers lack key ecological data, upon which predictive models depend. Research is needed to fill those gaps and the Comparative Analysis of Marine Ecosystem Organization (CAMEO) program has significant potential to do so. Finally, an ecosystem approach to management will require easily interpretable products to help integrate and convey complex ecosystem information to managers. Integrated Ecosystem Assessments (IEAs) will facilitate this information transfer, and it is recommended that IEAs be developed on both regional (large marine ecosystem) and sub-regional scales.

Regarding objective 2, broader stakeholder participation can be most effectively incorporated by expanding membership on Council committees to include non-fishing interests, and by increasing methods of communication among stakeholder groups and between these groups and the Councils. This broader stakeholder participation is needed to ensure that a more comprehensive ecosystem perspective is considered. It is also recommended that the rotation of Council meeting locations may help ease the cost and logistical burdens of stakeholder attendance and therefore encourage more stakeholders to participate. Stakeholder surveys should also be expanded to help ensure that a range of non- fishing views are considered in the fisheries management process. Previous survey results have demonstrated that stakeholders value ecosystem goods and services beyond fisheries and have resource use patterns that are important to consider in the management process. Finally, the level of interagency communication must be increased. Multiple agencies have jurisdiction over various ecosystem components, and it is crucial for these agencies to communicate ecosystem knowledge with each other and coordinate their management actions from a holistic and integrated ecosystem perspective. To achieve this, these agencies must be involved in the Council processes.

Regarding objective 3, processes to account for the effects of environmental variation on fish stocks and fisheries must consider climate-scale variability. Climate change is a growing concern for marine ecosystem management, and climate-induced change is inevitable. Thus, it is recommended that collaborations with climate change researchers be maintained and that climate and ecosystems modeling efforts be strengthened. Simultaneously, efforts to better understand the role of non-climate, human-induced changes in coastal systems on fish populations also must be pursued with added vigor. Burgeoning populations in U.S. areas adjacent to coasts, estuaries, and rivers directly and indirectly impact habitats vital to harvested fish and their forage.

It is recommended that management strategy evaluations be used to determine the appropriate environmental variables necessary to improve stock assessment performance. Incorporating environmental variables or indices into stock assessments is not an easy task, and environmental indices need to be very reliable in order to offset the potential risk associated with erroneous predictions. Management strategy evaluations are a means of determining which indices are beneficial. To improve predictive models it is also recommended that there be a focus on understanding critical mechanisms underlying correlations between environmental variability and fish productivity. Programs such as FATE (Fisheries and the Environment) and NPCREP (North Pacific Climate Regimes and Ecosystem Productivity) are examples of programs aimed at such integration. Environmental variations can influence physiological conditions, such as growth and reproduction, and finer spatial and temporal scale sampling may help to clarify these relationships. Finally, while multi-species and ecosystem models are being developed and improved, there is a need to maintain conventional single-species stock assessments, as there is no indication these new models will reduce the need for conventional models in the near term.

Regarding objective 4, existing and developing Council efforts to implement ecosystem approaches vary by region and many challenges remain. While some Councils are actively moving forward with ecosystem approaches to management (EAM), others are awaiting more definitive national guidance. Efforts include establishing Fishery Ecosystem Plans (FEPs), holding public meetings and workshops to discuss EAM, conducting ecosystem user surveys, developing ecosystem models, and mapping essential fish habitat. It is also important to recognize that many existing Council management efforts (such as bycatch reduction, area closures, and fishing fleet reduction) already represent significant progress toward an ecosystem approach. Despite this progress, additional effort is required. One challenge is the lack of sustained, annual support for FEP development and implementation. Another challenge concerns the complex jurisdictional environment in which ecosystem components are managed. Multiple federal, state, and local agencies have authority over different aspects of the ecosystem, and these roles need to be further defined and coordinated at the agency, inter-agency, and Council levels. Councils should consider ways to conduct more extensive outreach with these other entities to better incorporate their input into the Council process. Similarly, other entities should look for opportunities to engage the Councils, where appropriate, in their processes and issues. This will help to improve inter-agency communication and support a broad ecosystem perspective.

INTRODUCTION

In 1996, NOAA's National Marine Fisheries Service (NMFS) established the Ecosystem Principles Advisory Panel (EPAP) pursuant to Section 406 of the Magnuson-Stevens Fishery Conservation and Management Act. The panel was charged to submit a report to Congress that included: 1) an analysis of the extent to which ecosystem principles are being applied in fishery conservation and management activities, including research activities; 2) proposed actions by the Secretary and by the Congress that should be undertaken to expand the application of the ecosystem principles in fishery conservation and management; and 3) other such information as appropriate. The panel's report, submitted to Congress in 1999, outlined a number of principles, policies, and goals for ecosystem-based fisheries management as well as recommendations for practical steps that could be taken to implement an ecosystem approach (NMFS, 1999).

Now, 10 years later, Section 406 of the reauthorized Magnuson-Stevens Act charges NMFS, in consultation with the Fishery Management Councils, to undertake a study on the "state of the science for advancing the concepts and integration of ecosystem considerations in regional fishery management." To gather information for this study, a workshop was held at the NOAA Alaska Fisheries Science Center in Seattle on January 9–10, 2008. Workshop participants included a mix of ecosystem scientists from the NMFS Science Centers, NMFS Regional Offices, Councils, and some former members of the EPAP. This Report to Congress is based on the results of this workshop. Its structure speaks to the four overall objectives of Section 406 of the 2007 MSFCMA.

1. RECOMMENDATIONS FOR SCIENTIFIC DATA, INFORMATION, AND TECHNOLOGY REQUIREMENTS FOR UNDERSTANDING ECOSYSTEM PROCESSES, AND METHODS FOR INTEGRATING SUCH INFORMATION FROM A VARIETY OF FEDERAL, STATE, AND REGIONAL SOURCES

Recommendations in the 1999 EPAP Report addressed scientific data, information, and technology requirements for understanding ecosystem processes. These recommendations included a call for a description of long-term data and how these data are used for monitoring and management, a description of habitat needs for different life history stages of living marine resources, development of a broad suite of indicators to monitor ecosystem status, and accountability for total fishery removals (NMFS, 1999). A number of efforts are underway to provide such information. NMFS has published a Data Acquisition Plan that details the routine fisheries-dependent and fisheries- independent data required to support U.S. federal marine fisheries management (NMFS, 1998). Additionally, programs such as the Integrated Ocean Observing System (IOOS) are prioritizing new data collections and providing the framework for improving access to those data. A new program, Comparative Analysis of Marine Ecosystem Organization (CAMEO), focuses on improving the science support for ecosystem management by uncovering how the structure and function of ecosystems respond to change, human impacts, and environmental variation. Such programs will help to provide some of the data called for by the EPAP, but additional data are needed for a full understanding of ecosystem processes. Some of the major data and information requirements, and recommendations to address them, are detailed below.

The Role of Surveys, Assessments, and Other Routine Data Collection Efforts and Analyses in Understanding Marine Ecosystems

Surveys, assessments, and other routine data collection efforts and analyses play an important role in understanding marine ecosystems. Currently, fishery-independent surveys provide vital information on the relative abundance and ecological characteristics of many fish and invertebrate stocks as well as non-fishery species (such as marine birds, turtles, and mammals) in U.S. marine ecosystems. Zooplankton and ichthyoplankton surveys provide long-term time series data for early life history stage abundances of exploited and unexploited species. Programs such as the California Cooperative Oceanic Fisheries Investigations (CALCOFI) surveys off California, Marine Resources Monitoring Assessment and Prediction (MARMAP) and Continuous Plankton Recorder surveys in New England, and Fisheries-Oceanography Coordinated Investigations (FOCI) surveys in Alaska provide the basis to monitor secondary production and fish recruitment dynamics in those regions and provide essential ecosystem information. Fishery-dependent sampling provides key information on the effects of fishing on exploited and unexploited species and includes collecting landings data from commercial and recreational fisheries, logbook reporting, and vessel monitoring systems. In part as a response to the EPAP, a comprehensive Observer Program has been implemented in all regions of the United States to monitor the amount of catch discarded at sea so total removals of fished species can be calculated. Each of these survey types provides the basis for long-term time series of information critical for monitoring of ecosystems.

Despite the useful information provided by such surveys, notable gaps still exist in necessary scientific data. Most surveys target exploited open ocean species and species that co-occur with such exploited species. As a result, some fish stocks, especially those in nearshore habitats and their forage, remain virtually unsampled. An unrelated issue pertains to survey frequency. In some regions, surveys are conducted only every third year, which does not provide the necessary data to detect annual fluctuations. Additionally, lack of time-series data on the benthos and the failure to understand the detailed topography of fishing areas, how exploited species (all life history stages) are distributed relative to such features, and how fisheries species and their forage gain benefit from their habitats are daunting impediments to understanding relationships between habitat, benthic fauna, and fish species. It has limited the ability to determine the effects of mobile fishing gear on habitats and to fully appreciate both the benefits and impacts of Marine Protected Areas (MPAs). Knowledge about the benthos is also important for understanding energy flow and food web dynamics, including benthic-pelagic coupling and benthos as prey of commercially important species. Expansion and development of such surveys will also provide useful data on biodiversity, a fundamental property of ecosystem structure. Another notable gap is in socioeconomic collections, which are a key component of ecosystem analysis and management. Understanding the behavior of harvesters and predicting their responses to changes in management measures and associated costs is critical.

Research and Technology Requirements Needed to Understand Critical Ecosystem Processes

Marine ecosystem management can be improved through research that elucidates underlying dynamics at a variety of scales and increases understanding of critical ecosystem processes. A critical area of research is the development of a variety of modeling methods, such as MPA site-selection models, statistical models, and/ or theoretical models. In most cases, modelers lack important ecological information, and research is required to fill these gaps. For MPA site-selection models, appropriate distribution and abundance data on most ecological groups are lacking. For these models, intensive small-scale surveys of all ecosystem components, ranging from physical conditions to marine mammals, and socioeconomic data on the human drivers and management impacts are necessary. Statistical models will improve understanding of links between habitat and demographic rates of marine organisms, functional feeding responses of predators, strength of ecological interaction, biodiversity, and connectivity among local populations, but to improve these data-driven models, process-oriented studies of ecological interactions and the increased use of lab, small-scale field, and management experiments are needed. Theoretical models are important for the development of ecological forecasts and IEAs but further information on ecological mechanisms is needed. For example, robust data on variation in demographic rates, functional responses, mechanisms limiting or generating change in population size, ecological interactions, and the combined effects of density-dependent and density-independent processes are needed.

This ecological modeling effort will provide a greater basic understanding of ecosystem processes and practical tools for evaluating the effectiveness of ecosystem-based management efforts, but there is also a need for a variety of technologies required to efficiently expand data collections to meet these needs. New sensor capabilities are necessary to realize the full potential of *in situ* ocean observing networks and satellite-based observations in implementing EAM. Technological advances for rapid measurement and integration of surface and subsurface chlorophyll concentrations are needed for improving estimates of sustainable carrying capacity of ecosystem goods and services. Moored and autonomous instruments can vastly increase spatial and temporal resolution of ecosystem monitoring data. Further development and adaptation of Automated Underwater Vehicles (AUVs), Remote Underwater Vehicles (ROVs), autonomous underwater gliders, and vehicle types yet to be described, specifically for deployment of EAM advanced sensors, are also needed. There is also a need to increase the capacity to sample adaptively. Critical events affecting recruitment may be short-lived and/or localized. This requires flexible vessel resources and instruments designed to sample when a trigger threshold is crossed. Programmable drifters with behavior and trigger released drifters to sample eddies are being developed and can increase sampling of key oceanographic features at a relatively low cost. Increased deployment of autonomous smart instrumentation can supplement and extend coverage by ship-based surveys. Continued support of research to develop these innovative sensors is critical. Remote sensing data are available from the U.S. National Aeronautic and Space Administration, as well as NOAA, and the continuation and development of remote sensing capabilities is essential. Continuity of high-resolution, reliable, remote sensing data (sea surface temperature, sea surface height, ocean color) under the next generation of sensors is fundamental for EAM.

Integrating Ecosystem Information from Various Sources to Improve Management

There is an ongoing need to integrate the ecosystem information collected by multiple institutions to improve ecosystem management. Partnerships with state and local government agencies, research institutions, and universities will be the key to integrating information effectively. Programs, such as IOOS, have been initiated to establish an integrated and sustained coastal ocean observation system capable of meeting diverse regional and national information requirements for the purposes of advancing ocean science and resource management. The focus of NOAA's IOOS plan is to improve access to high-quality, integrated data, and to enhance data products and decision support tools (National Ocean Service, 2007). Support of integrated information systems, such as IOOS, is essential to the development of the data needed to conduct EAM.

The current implementation of IOOS and related observational efforts needs to be augmented to focus more fully on environmental/habitat conditions in nearshore waters, estuaries, and coastal rivers; changes in land use and land cover; alterations in freshwater delivery to coastal systems via rivers and groundwater; and atmospheric delivery of substances that affect ecosystem functioning and the ability of coastal systems to sustain harvestable populations of fish and shellfish. Burgeoning populations in U.S. areas adjacent to coasts directly and indirectly impact habitats vital to harvested fish and their forage. The resulting alterations frequently are regional in scale and are capable of inducing changes as profound in intensity as climate alteration.

There is also a growing need to integrate and deliver ecosystem information to managers and stakeholders in a broad array of products (websites, reports, briefings) to facilitate information transfer. Novel ways of distilling information into easily interpretable products are important. The development of such information products (e.g., IEAs) is the key to addressing the needs of data integration and production of useful products to decision-makers and stakeholders. An IEA is a formal synthesis and quantitative analysis of information on relevant natural and socioeconomic factors in relation to specified ecosystem management goals (Levin et al., 2008). IEAs involve and inform citizens, scientists, managers, and policymakers through formal processes to contribute to attaining the goals of EAM. IEAs should be developed on a regional and sub-regional scale. They will provide a framework into which various forms and sources of ecosystem data can be integrated on a large marine ecosystem scale.

Recommendations

- Maintain and expand current levels of fishery-dependent and fishery-independent data collection activities.

- Expand surveys collecting benthos and habitat-related data (e.g., area of coverage, scale of data collection, and focus on habitat use by resource and forage organisms) and increase socioeconomic data collection efforts.
- Accelerate the development of ecosystem models. The CAMEO program has the greatest potential to contribute to the development and improvement of experimental and operational ecosystem models.
- Develop and implement the next generation of sensors and undersea vehicles to fulfill the increased data needs for EAM. Deployment of innovative technologies such as autonomous smart instrumentation will supplement and extend coverage by ship-based surveys. Flexible and increased vessel resources will also be needed to allow adaptive sampling as well as the intensive sampling needed for understanding processes and rates. Continuity of high-resolution, reliable, remote sensing data also needs to be maintained.
- Support the continued development of integrated information systems, such as IOOS, which are essential for the provision of data needed to conduct EAM.
- Enhance data collection and analyses in near coastal, estuarine, and riverine systems to capture effects of coastal population growth on fishery production.
- Develop and implement integrated ecosystem assessments (IEAs) on regional (large marine ecosystem) and sub-regional scales to provide a practical framework for the integration of ecosystem data and a means to make informed ecosystem-based management decisions.

2. RECOMMENDATIONS FOR PROCESSES FOR INCORPORATING BROAD STAKEHOLDER PARTICIPATION

The 1999 EPAP report recommendations addressed the need for broad stakeholder participation in effective ecosystem approaches to management. Effective EAM relies on the participation, understanding, and support of a broader suite of stakeholders than that currently participating in the fishery management process. In particular, the EPAP noted that effective EAM would require including stakeholder groups indirectly affected by fisheries as well as industry sectors that indirectly affect fisheries (e.g., through water quality). The full participation of all stakeholders, including the interests of future generations, was noted by the EPAP as likely to result in policy development and implementation that is more fair and equitable.

Many of the Councils have conducted workshops and outreach activities to educate and inform constitutients on the ecosystem approach to fisheries management. Surveys have also been implemented to gather public input on a variety of ecosystem-related issues. Such actions are designed to reach beyond the fishing industry to incorporate a broader range of stakeholders into the EAM process.

Current stakeholder involvement and mechanisms for broadening stakeholder participation in an EAM are outlined below.

Major Ecosystem Stakeholders Already Participating in the Fisheries Management Process and Those Missing from the Process

EAM, by its nature, requires participation of a broad range of stakeholder groups to provide for a more holistic ecosystem perspective. This effort must move beyond fisheries interests alone to adequately address all emerging management concerns and efforts. One possible vehicle for this expansion is the Council process, which may allow for increased participation by state and federal representatives. In the current advisory capacity of the Council process, there are meetings, hearings, science and statistical committees (SSCs), advisory panels (APs), and other committees. These processes can be broadened through the expansion of current bodies to enhance representation, creation of new committees, or development of ecosystem advisory panels.

Some but not all ecosystem stakeholders are currently participating in the fishery management process. Organizations such as coastal groups, coastal developers, regional watershed resource managers, and energy companies/ organizations may have an interest in Council activities and/or may be undertaking activities that affect fisheries. For example, with the exception of the Coast Guard, the U.S. military is often missing from the process even though their activities can have ecosystem effects and they can be an important source of oceanographic and hydrographic information. The U.S. Geological Survey (USGS) and Minerals Management Service (MMS) are also important agencies to involve based on their data/scientific expertise (USGS) or their management goals (MMS), which might be in conflict with fishery management goals. Currently, the representation of such agencies on Councils varies regionally. Even the participation of mandated organizations, such as the U.S. Fish and Wildlife Service, varies regionally according to the specific management concerns being addressed. Successful EBM will necessitate two-way communication between the Councils and this broader base of stakeholders. It will be important for the Councils to seek and be given the opportunity to provide input into the management activities of these stakeholders as well. The Councils should seek to establish additional forums to engage various other federal and state agencies as they pursue their legislative mandates in order to broaden coordination and align management and science programs.

Effective Means of Bringing Stakeholders and Their Interests into the Current Management Council Process to Provide a Broader Ecosystem Perspective

Several means exist to incorporate a broader stakeholder base into the current management council process. An improved system of outreach and notification to pertinent stakeholders could encourage increased participation. Current means of public notification and communication could be expanded to include these other stakeholder groups to ensure that they are aware of the schedule and agendas for Council meetings and/or ecosystem advisory group meetings and of the various ways they can participate in the Council process. Similarly, other stakeholders and management entities should look for opportunities to engage the Councils in their processes and issues. Council meetings are generally rotated among various locations, although the locations may be limited due to cost or other logistical issues. The rotation of meeting locations allows participation of more groups by relieving some of the financial burden associated with traveling to meetings.

Statistically based surveys of the general public and/or more targeted studies of particular stakeholder groups may serve as another means of increasing stakeholder participation by evaluating societal values and resource use patterns. A number of these surveys have recently been implemented to obtain information on ecosystem attitudes, preferences for management options, and valuation surveys on MPAs, protected species, and corals. These have included non-market valuation surveys of public willingness to pay for Steller sea lion recovery (NMFS, 2007) and studies that have looked at the current Council composition and voting relative to various issues (Ellis, 2008). Results have demonstrated that the public values marine resources and has preferences on use patterns.

NOAA's vision for EAM specifically includes the need to consider stakeholder requirements in science and management. Information-sharing among affected agencies may assist managers in planning and choosing among management options that affect various stakeholder groups. There is a significant need for an overall increase in the level of inter-agency communication and collaboration to help decision-makers gain a holistic perspective, but some Councils have expressed frustration with the lack of communication. Increased communication may need to be accomplished through establishment of formal structures and may need high-level intervention to facilitate cooperation among agencies. In Alaska, there are some efforts to implement this recommendation through the Ecosystem Committee of the North Pacific Fishery Management Council, the Alaska Marine Ecosystem Forum, and fishery ecosystem plans that involve multiple agencies in development and implementation (NPFMC, 2008).

Recommendations

- As appropriate, expand Council APs, SSCs, and other committees to include relevant ecosystem stakeholders and/or create new ecosystem committees or advisory panels.
- Develop methods for communicating with stakeholders, including rotating Council meetings or Council committee meetings among coastal communities. Increased communication will encourage broader participation and the rotation of meeting locations will ease the logistical burdens of stakeholders interested in participating.
- Increase surveys of stakeholders and the general public (consider cross-agency surveys and other information sharing) to provide information on and increase knowledge of societal values and resource use.
- Increase the level of agency-to-agency communication/collaboration. This may need to be done through establishment of formal structures and may need high-level intervention.

3. RECOMMENDATIONS FOR PROCESSES TO ACCOUNT FOR EFFECTS OF ENVIRONMENTAL VARIATION ON FISH STOCKS AND FISHERIES

Recommendations in the 1999 EPAP report focused mainly on general ecosystem issues, and were silent with regard to detailed modeling of environmental effects, especially in single- species assessments. Since that time, global climate change has emerged as a significant threat to marine ecosystems and fisheries, and will be an important consideration in EAM. NOAA is developing strategies for addressing the impact of climate on living marine resources and coastal ecosystems (Griffis et al., 2008). Consideration of the effects of environmental variability on fisheries will need to include an explicit emphasis on climate-scale variability and change. NOAA needs to maintain collaborative interactions with global climate change researchers, and climate and ecosystem models need to be strengthened.

The influence of environmental variability on fish stocks is an area of active scientific research. Over the past 10 years, the National Research Council, often commissioned by NMFS, has published a number of reports discussing the incorporation of environmental variability into stock assessments and the associated difficulties, indicating the importance and complexity of this topic (NRC, 1997, 1999, 2002, 2006). NMFS has also conducted a variety of workshops covering this topic (Mace, 2003; Watters, 2004; Methot, *in prep*). As climate change continues to threaten marine ecosystems, understanding relationships between environmental influences and fish productivity will continue to be important.

Some of the major challenges associated with accounting for environmental variability and the ability of current stock assessments to account for this variability are discussed below.

Major Challenges in Accounting for the Environmental Variability on Fish Stocks and Fisheries

Accounting for the impact of environmental variability on fish stocks is not an easy task, and many challenges still exist. Including environmental explanatory variables may potentially improve predictions, but this also allows greater scope for erroneous predictions, with a corresponding increase in risk. The few relevant simulation studies agree that an environmental variable must be a very reliable predictor for the benefit to outweigh the risk, and in practice this reliability standard can seldom be met (Basson, 1999; Kell et al., 2005). Simulation studies or "management strategy evaluations" need to be conducted to determine which environmental variables in stock assessments improve performance and which degrade performance.

Improvement of predictive models is dependent on increasing the current level of understanding of critical mechanisms influenced by environmental variability. For example, the capability to predict recruitment will require further understanding of environmental mechanisms underlying spatial and temporal fluctuations in fish production, combined with local ocean circulation models (e.g., Regional Ocean Modeling System models), and will most likely need to be done on a case-bycase basis. Additionally, there are environmental influences on physiological condition relating to growth, maturation, fecundity, and timing of reproduction, but these climate-related properties are seldom being monitored in annual fish catches or surveys. Increased ecological monitoring and modeling will potentially improve understanding of fluctuations in natural mortality rates. There is a need to maintain emphasis on understanding critical mechanisms, especially those underlying apparent correlations between environmental conditions and fish stock productivity as well as increased monitoring of annual fluctuations in growth and reproductive condition. Current programs, such as FATE (Fisheries and the Environment) and NPCREP (North Pacific Climate Regimes and Ecosystem Productivity), are examples of programs aimed to provide such critical information. Further understanding may require finer spatial and temporal resolution of physical and biological conditions. As researchers and managers consider the possible impacts of global climate change on marine ecosystems, it will become increasingly important to understand these relationships.

Ability of Existing Stock Assessments and/or Harvest Advice Processes to Adequately Account for the Effects of Environmental Variability on Fish Stocks and Fisheries

The present ability to account for environmental effects is minimal. It has been repeatedly shown that under conventional single-species management approaches, there is little benefit in anticipating year-to-year fluctuations unless those predictions are more reliable than can be achieved currently (Walters and Parma, 1996; Myers, 1998; Stokes et al., 1999; Patterson et al., 2001). Current multi-species models have very limited predictive accuracy, and ecosystem shifts can generally be recognized only after they have happened. It will be more beneficial to shift from focusing on inter-annual variability to low frequency or inter-decadal environmental variability and to understanding and better predicting the nature and intensity of habitat changes resulting from human activities. To improve understanding of inter-decadal variability, it will be important to continue to capture relevant existing historical datasets, as sources are discovered and opportunities arise. The NESDIS Climate Database Modernization Program, run by the National Climatic Data Center, has been performing an especially valuable service in this regard.

Although the importance of habitat is widely recognized, scientists lack the quantitative ability to model, assess, and anticipate changes in resource productivity resulting from habitat alteration, habitat loss, and habitat restoration. There is a need for further understanding of anthropogenic effects (e.g., degraded water quality, habitat loss, and water diversion) and for improving the quantitative basis for assessing habitat. Anthropogenic habitat alterations, such as salinity alterations associated with changes in regional water usage and enhanced sediment and nutrient loads caused by wide-scale changes in land use patterns, are capable of inducing changes as profound in intensity (although not in geographic scope) as climate alteration. Another area where much could be gained from better understanding of habitat factors and environmental effects is that of stock rebuilding, especially where there is reason to believe that the present ecosystem and habitat may be different from the historical conditions used to develop rebuilding targets and expected rebuilding times. In the worst cases, such as catastrophic loss of habitat (e.g., Sacramento winter run Chinook salmon), it may not be possible to rebuild a stock to historical levels, but it will be necessary to be able to provide a basis for an appropriate rebuilding target.

As NOAA continues to move forward with ecosystem-based management, it will be important to maintain single-species stock assessments. Although environmental and ecosystem relationships may have promise, such improvements can only be built upon a firm single-species foundation. A plan needs to be developed and implemented articulating the steps for moving from single-species to ecosystem-oriented management. In its 2008 evaluation of the state of the use of ecosystem models (Townsend et al. 2008), NMFS recommended an enhanced program to develop, test and implement promising new classes of such models.

Recommendations

- Maintain collaborative interactions with global climate change researchers and strengthen climate and ecosystem models.
- Conduct simulation studies or "management strategy evaluations" to determine when including environmental variables (including those pertaining to habitat) improves stock assessment performance.
- Maintain emphasis on understanding critical mechanisms, especially those underlying apparent correlations between environmental conditions and fish stock productivity. This may require measurements of physical and biological conditions at finer spatial and temporal scales.
- Increase routine monitoring of annual fluctuations in growth and reproductive condition. This is a mundane and low-tech aspect of fish stock monitoring with very high potential for improving understanding of environmental effects.
- Shift from focusing on interannual variability to focusing on low-frequency or interdecadal variability (i.e., long strings of good or bad years). This may include trends in physical and biological patterns as well as shifts at an ecosystem level.
- Enhance understanding of anthropogenic effects on habitat and fishery productivity. Anthropogenic effects may be more influential in some regions and fisheries than others, but increasingly are being seen as regional rather than local in scope. They represent some of the most promising opportunities for effective science-based anticipatory actions leading to management control.
- Maintain the emphasis on supplying adequate information for conventional single-species stock assessments. There is no indication that an ecosystem approach will reduce the ongoing need for conventional stock assessments in the near term.

4. Description of Exisiting and Developing Council Efforts to Implement Ecosystem Approaches, including Lessons Learned by the Councils

The 1999 EPAP report concluded that the eight Councils are implementing many ecosystem principles, but these principles are not applied comprehensively or evenly across Council jurisdictions. Below, the Councils' efforts, successes, and challenges associated with implementing EAM are discussed.

Current and Developing Council Efforts to Implement Ecosystem Approaches

The eight Councils are at differing stages of progress and have taken different paths in developing an EAM. All Councils recognize the mandate to incorporate ecosystem principles into fishery management, but lack the understanding and tools to do so fully. The necessary steps for implementing EAM are not always clear, and some Councils are not attempting formal implementation, but rather have proceeded cautiously, awaiting national guidance. At least two Councils, however, have actively moved forward with implementing EAM. The North Pacific Fishery Management Council (NPFMC) has established a Fishery Ecosystem Plan (FEP) for the Aleutian Islands (NPFMC, 2007); and the Western Pacific Regional Fishery Management Council (WPFMC) has formally established an ecosystem policy to develop FEPs according to a place-based approach (WPFMC, 2007).

Councils that have not implemented FEPs have taken a vareity of steps towards EAM. The Pacific Fishery Management Council (PFMC) has moved to begin development of an Ecosystem Fishery Management Plan that could be patterned after the NPFMC FEP and is envisioned to take the form of an umbrella plan that integrates ecosystem considerations across existing FMPs. PFMC has held multiple public Council sessions and advisory body meetings on the topic and is currently seeking funding to convene a plan development team.

The New England Fishery Mangament Council (NEFMC), Mid-Atlantic Fishery Management Council (MAFMC), South Atlantic Fishery Management Council (SAFMC), and the Gulf of Mexico Fishery Management Council (GMFMC) used 2004 congressional funding to solicit constituent views about EAM and to support local ecosystem-related projects. The NEFMC conducted public meetings, workshops, and surveys on fishery values, management objectives, and tradeoffs between potential loss of total allowable catch relative to allocations for ecosystem needs. The MAFMC conducted educational workshops and activities to acquaint Council members, constituents, and the general public with the new emphasis on EAM and discuss options relative to implementation. Such education and outreach efforts have proven to be beneficial. Other Councils have used the funds to focus on modeling and field efforts to support EAM. The SAFMC has begun the development of an FEP using EPAP guidance and, like the GMFMC, has been exploring fisheries-based ecosystem modeling and other trophic models. The Caribbean Fishery Management Council (CFMC) conceived an ecological approach to fishery management as early as the 1980s, but funding limitations and FMP requirements did not allow for development of FEPs at that time. The CFMC has since concentrated on mapping and characterization of essential fish habitats (EFH), and is currently exploring ecosysytem modeling for fisheries management.

Successful Council Steps toward EAM Implementation

The 1996 Magnuson Stevens Act requirements, including the National Standards, and other requirements such as the National Environmental Policy Act (NEPA) and the essential fish habitat provisions, have resulted in fishery management actions that can be considered progress toward the implementation of EAM. In addressing requirements relative to bycatch, overfishing, protected species, EFH, and cumulative effects assessment, the Councils have implemented gear restrictions, time and area closures to protect spawning populations, conservative and environmentally influenced harvest policies for forage species. These policies include harvest prohibitions, protection of habitat areas of particular concern (HAPCs), catch quotas, and bycatch and fishing effort restrictions. Some Councils have become involved in understanding and protecting non- fishery species, such as corals, marine mammals, sea turtles, and seabirds. All of these actions have indirect, if not direct, EAM implications. Councils cite success in the ending of overfishing of certain species, reduction of fleet capacity, and moving from single species to multi-species management as moving toward EAM. Mapping and characterization of habitats by Councils and development of EFH amendments to the FMPs also exemplifies the ecosystem approach.

Councils have taken different approaches toward ecosystem mangement and establishment of an ecosystem policy; FEPs by the WPFMC and NPFMC provide examples. The WPFMC is transitioning from FMPs to FEPs. The new FEPs are place-based in consideration of differing local conditions and cultures. The NPFMC has established an FEP for the Aleutian Islands without replacing existing FMPs. The FEP is comprehensive in providing the best available science on understanding the ecosystem as a framework within which to make fishery management decisions. The PFMC is in the early stages of developing a similar umbrella type plan. These two differing approaches offer potentially viable options for the other Councils.

Challenges Associated with Ecosystem Implementation

EAM is a more complex and information-intensive process than traditional fisheries management approaches and will require dedicated resources to implement effectively. At present, Councils cannot undertake EAM as a dedicated programmatic task and NMFS is unable to provide the required environmental and fisheries data and associated predictive analyses.

Another major challenge to implementing EAM relates to jurisdiction and authority. Many ecosystem attributes, such as water quality, are impacted by activities regulated by other agencies. While Councils can certainly include such ecosystems components in their management considerations, they currently do not have authority over many important ecosystem components. If the Councils' responsibility is extended to encompass ecosystem issues beyond target fishery species and their habitat, these jurisdictional issues need to be addressed. Legislative change that grants additional jurisdictional authority may be needed to allow the Councils to manage more broadly. Furthermore, most Councils prioritize their responsibility under the Magnuson-Stevens Act to achieve fisheries harvest at optimum yield. Finding a balance between achieving a beneficial fishery harvest and protecting marine ecosystems continues to be a challenge. Jurisdictional authority needs to be redefined at the agency, Council, state, and national level.

In summary, common EAM issues exist across the Councils. Foremost among these issues is the lack of clarity among Councils as to how to implement EAM. The guidelines for implementation are ambiguous, and the role of the Councils in the context of ecosystem-based management is not defined. Moreover, the additional demands of EAM on staff and the funding limitations are major impediments toward progress on EAM and FEP within NMFS and the Councils. The Councils recognize that good science is crucial to providing the information and advice that will help address EAM, but the lack of dedicated funds to support the engagement of the scientific community is a problem.

Recommendations

- Provide each of the Councils with sustained, annual funding to develop and implement FEPs.
- Keep science at the forefront of the process in developing FEPs. The Councils strongly favor having the "best available science" to understand ecosystem attributes for making management decisions.
- Provide a better understanding of what constitutes an EAM, including more definitive and detailed guidance to Councils on how to develop FEPs.
- Develop FEPs incrementally, building upon each Council's past and current stage of progress moving toward EAM. Determination of what constitutes a successful FEP needs to be established.
- Clarify and redefine jurisdictional authority at multiple levels: 1) within NOAA, 2) within the Federal Government (both agencies and legislative mandates), 3) at the Council level and 4) with respect to the jurisdiction of states and other local authorities. Additional legislative authority may be needed if the Councils are to expand their role in EAM beyond fisheries and EFH.
- Increase opportunities for and develop Council processes, as appropriate, to increase and incorporate input from other federal agencies (such as the Department of the Interior and Environmental Protection Agency) and stakeholders with ecosystem interests.

SUMMARY

NOAA and the Councils are continuing to incorporate ecosystem principles into fishery management efforts. As charged by Section 406 of the reauthorized Magnuson-Stevens Act, this report examines the "state of the science for advancing the concepts and integration of ecosystem con-siderations in regional fishery management." A variety of recommendations will need to be addressed to successfully and fully implement an ecosystem approach to management.

While an ecosystem approach to management does not require a complete understanding of each ecosystem, it does require fundamental knowledge of basic ecosystem principles. These principles, as outlined by the EPAP (NMFS, 1999), highlight the complex and dynamic nature of ecosystems and researchers' limited ability to predict change. With these principles in mind, scientific data, information, and technology requirements necessary to further understand ecological processes have been identified. This will require sustained ecosystem observations, process-oriented research, and integrative modeling in support of an ecosystem approach to management. Specific biological, socioeconomic, and climate/physical data needs should be addressed to strengthen EAM. The CAMEO program was created to strengthen the fundamental scientific basis for an ecosystem approach to management and will be an important ingredient to moving forward.

An ecosystem approach to management requires broad stakeholder involvement. Specifically, it should include the participation of stakeholders interested in the ecosystem and effects on the ecosystem, beyond fishing alone. These stakeholders should be engaged in the science and management of the resources. Reaching more stakeholders can be done through a variety of means, such as improving communication methods, outreach, and agency-to-agency collaborations. EAM represents a more holistic approach to management and, consequently, a more holistic group of stakeholders should be identified.

An ecosystem approach to management should also seek to account for the effects of environmental variation on fish stocks and fisheries. Global climate change is a major concern and understanding environmental effects will become increasingly important. Emphasis should be maintained on understanding mechanisms underlying apparent correlations between environmental conditions and fish stock productivity as well as anthropogenic effects. Environmental variation is inherently difficult to predict, but increased monitoring and models will help to improve the understanding of environmental effects. Incorporating this knowledge into EAM will not reduce the need for conventional stock assessments, however, because future improvements will need to be built on a solid single-species foundation.

These recommendations will help guide NMFS as it strives to improve management policies. The waters within U.S. Government jurisdiction are home to valuable living resources that must be protected and conserved. These resources provide substantial economic, social, and cultural benefits. Currently, many of these resources are harvested at high levels, which necessitates conscientious management decisions.

REFERENCES

[1]   Basson, M. (1999). The importance of environmental factors in the design of management procedures. *ICES J. Mar. Sci, 56*, 933-942.

[2]   Ellis, B. (2008). Conflict of interest standards and regional fishery management councils: An evaluation of the North Pacific Council's voting record on conservation issues, 14 p. Institute of the North, Anchorage, Alaska.

[3]   Griffis, R. B., Feldman, R. L., Beller-Simms, N. K., Osgood, K. E. & Cyr (editors), N. (2008). *Incorporating Climate Change into NOAA's Stewardship Responsibilities for Living Marine Resources and Coastal Ecosystems: A Strategy for Progress.* U.S. Dep. Commerce, NOAA Tech. Memo. NMFS-F/SPO-95, 89 p.

[4]   Kell, L.T., Pilling, G. M. & O'Brien, C. M. (2005). Implications of climate change for the management of North Sea cod. *ICES J. Mar. Sci, 62*, 1483-1491.

[5]   Levin, P. S., Fogarty, M. J., Matlock, G. C. & Ernst, M. (2008). Integrated ecosystem assessments. U.S. Dep. Commer. NOAA Tech. Memo. NMFSNWFSC-92, 20 p.

[6]   Mace & Pamela M. (Ed.). (2003). Proceedings of the Seventh NMFS National Stock Assessment Workshop. NOAA Tech. Memo. NMFS-F/SPO-62, 46 p.

[7]   Methot & Richard D. (Ed.). *In prep.* Proceedings of the Eighth NMFS National Stock Assessment Workshop. Contact *<Richard.Methot @noaa.gov>.*

[8]   Myers R. A. (1998). When do environment-recruitment correlations work? *Rev. Fish. Biol. Fish, 8*, 285-305.

[9]   National Marine Fisheries Service (NMFS). (1998). *NOAA Fisheries Data Acquisition Plan.* Retrieved on October 29, 2008 from http://www.st.nmfs.noaa.gov/st4/documents/DataPlan.pdf
1999. Ecosystem-based fishery management: A Report to Congress by the Ecosystem Principles Advisory Panel, 54 p. U.S. Dep. Commer., NOAA, NMFS, Silver Spring, MD.
2007. *Steller sea lion study: summary of survey results.* Retrieved on October 29, 2008 from http://www.afsc.noaa.gov/REFM/ Socio economics/ Projects/NCVSSLPM.php

[10]   National Ocean Service. (2007). *NOAA Integrated Ocean Observing System (IOOS) Program.* Retrieved on October 24, 2008 from http://ioos.noaa.gov/ library/ioos proj statplan.pdf.

National Research Council (NRC). (1997). *The Global Ocean Observing System: Users, benefits, and principles.* National Academy Press, Washington, DC. 1999. *Sustaining marine fisheries.* National Academy Press, Washington, DC.
2002. *Science and its role in the National Marine Fisheries Service.* National Academy Press, Washington, DC.
2006. *Dynamic changes in marine ecosystems: Fishing, food webs, and future options.* National Academy Press, Washington, DC.

[11] North Pacific Fishery Management Council (NPFMC). (2007). Aleutian Islands Fishery Ecosystem Plan – December 2007. North Pacific Fishery Management Council, 605 West 4th, Suite 306, Anchorage, AK 99501-2252 (http://www.fakr.noaa.gov/npfmc/contact.htm)
2008. *Overview of the Aleutian Islands Fishery Ecosystem Plan.* 22 p. North Pacific Fishery Management Council, Anchorage, Alaska.

[12] Patterson, K. R., Cook, R., Darby, C., Gavaris, S., Kell, L., Lewy, P. Mesnil, B., Punt, A., Restrepo, V., Skagen, D. W. & Stefánsson, G. (2001). Estimating uncertainty in fish stock assessment and forecasting. *Fish and Fisheries, 2,* 125−157.

[13] Stokes, T. K., Butterworth, D. S., Stephenson, R. L. & Payne, A. I. L. (1999). Confronting uncertainty in the evaluation and implementation of fisheries- management systems. *ICES J. Mar. Sci, 56,* 795–796.

[14] Townsend, H. M., Link, J. S., Osgood, K. E., Gedamke, T., Watters, G. M., Polovina, J. J., Levin, P. S., Cyr, N. & Aydin, K. Y. (editors). (2008). National Marine Fisheries Service Report of the National Ecosystem Modeling Workshop (NEMoW). U.S. Dep. Commerce, NOAA Tech. Memo. NMFS-F/SPO-87, 93 p.

[15] Western Pacific Regional Fishery Management Council (WPRFMC). (2007). Report on the 2007 Ecosystem Policy Workshop – November 2007. Western Pacific Fishery Management Council, 1164 Bishop Street 1400, Honolulu, HI 96813 (http://www.wpcouncil.org/contact.html)

[16] Walters, C. & Parma, A. M. (1996). Fixed exploitation rate strategies for coping with effects of climate change. *Can. J. Fish. Aquat. Sci, 53,* 148-158.

[17] Watters, George M. (Ed.). (2004). *Proceedings of the NMFS Workshop on Building Environmentally Explicit Stock Assessments.* National Oceanographic and Atmospheric Administration, National Marine Fisheries Service, Southwest Fisheries Science Center, Pacific Fisheries Environmental Laboratory. ADMIN REP PFEL no.04-01.

In: Ecosystem Assessment for Marine Resource...   ISBN: 978-1-61470-805-6
Editors: Maria P. Toscano   © 2012 Nova Science Publishers, Inc.

Chapter 3

VISION 2020: THE FUTURE OF U.S. MARINE FISHERIES

National Oceanic and Atmospheric Administration

FOREWORD

The Marine Fisheries Advisory Committee (MAFAC) advises the Secretary of Commerce on all living marine resource matters under the purview of the Department of Commerce. MAFAC members evaluate and assess national programs, recommend priorities, and provide their views on future directions. MAFAC members have a wide range of expertise, including but not limited to, commercial and recreational fishing, aquaculture, seafood processing, seafood marketing and sales, consumer interests, coastal communities, and environmental advocacy. MAFAC was established in 1970 to serve as a federal advisory body, complying fully with the Federal Advisory Committee Act.

INTRODUCTION

In September 2006, the Assistant Administrator of NOAA's National Marine Fisheries Service (NMFS) asked the Marine Fisheries Advisory Committee (MAFAC) to prepare a report on the desired future state of U.S. Marine Fisheries. The specific request from the Assistant Administrator to

MAFAC was "...to create, in clear, simple, non-jargon language, a stakeholders' consensus on the desired future state of domestic and international fisheries." This report is MAFAC's response. It is organized into three sections:

Section 1: Trends and their Impact on Marine Fisheries that provides context and reference points for comparison with the future;

Section 2: MAFAC Findings based on these trends; and

Section 3: SummaryRecommendations regarding fulfillment of MAFAC's vision of the future of our Nation's marine fisheries.

Appendices detailing the rationale behind the recommendations complete the report.

SECTION 1. TRENDS AND THEIR IMPACT ON MARINE FISHERIES

Marine fisheries have been, are, and will continue to be important to our Nation for a multitude of reasons. Marine fisheries provide employment and recreational opportunities as well as a food source. The passage of the Fishery Conservation and Management Act (FCMA) of 1976, P.L. 94-265,[1] (renamed in 1980 for the late Senator Warren Magnuson and in 1996 to include Senator Ted Stevens) and the establishment in 1983 of the exclusive economic zone (EEZ) ushered in a new era of federal fishery management. The United States has the largest EEZ in the world, 3.4 million square nautical miles. In addition, the United States' EEZ has a tremendous variety of fish stocks (in excess of 905 stocks[2]) and other living marine resources.

The dynamics of marine fish populations are affected indirectly by climate change, habitat availability, and water quality. They are also affected directly by human factors such as fishing and environmental degradation. Human fishing practices are affected by the dynamics of the marine ecosystem and fluctuations in fish abundance. Thus, a complex relationship exists between fish and fishermen that must be maintained to foster the existence of both. At the intersection of these complex interactions are fisheries managers who require high-quality observations and well supported predictions about species status and abundance. Accurate and precise biological, economic and social science data is required for management decisions. Presently, concerns arise if the biological, physical, social and economic data are deemed insufficient for managing marine fisheries sustainably.[3] The goal of fisheries management is to assure sustainable marine fisheries. In the simplest sense, sustainable use of a resource means that the resource can be used indefinitely.

Trend: Based on Status of Stocks Assessments, Global Fisheries Production Will Most likely Grow Slowly, If At All, to 2020.

Most assessments on the world-wide status of marine fisheries indicate that on a species by species level, most species considered have reached or are near maximum sustainable exploitation levels[4]. Thus, wild marine fisheries harvest which has peaked, at approximately 93 million tons per year on a worldwide basis[5], should not be expected to grow significantly.

Trend: The Consumer Demand for Fish and Shellfish Continues to Grow.

At the same time that marine fisheries harvest has plateaued or peaked, global consumption of fish has doubled since 1973[6]. Countries with rapid population growth, rapid income growth and rapid urbanization tend to have the largest increase in consumption of animal products including fish. The developing world has seen such increases. Today, fish and shellfish on average provide 25 percent of protein consumption in developing countries and 13 percent in developed countries. China, where income growth and urbanization are major factors, dominates consumption of fish products.

Trend: Seafood Consumption Is Increasing in the U.S. on a per Capita Basis

In 2006, Americans consumed 16.5 pounds (edible weight) per person, up from 16.2 pounds per person in 2005 and 0.9 lb higher than the 10-year average. Records were set in 2006 for per capita consumption of fillets and steaks, and shrimp in all forms of preparation.[7]

Trend: Consumption, Domestic and Worldwide, Is Expected to Increase as the Health Benefits of a Diet Rich in Seafood Protein Become Increasingly Recognized.[8]

This trend of a rising demand for seafood was recently confirmed by a panel at the annual meeting of the American Association for the Advancement of Science (AAAS). The panel further noted that demand will continue to exceed wild capture fisheries' ability to provide the fish meals demanded by consumers.

Trend: Although Domestic Wild-Catch Fish Stocks Are Improving, Domestic Demand for Safe[9] Seafood Will Continue to Exceed Domestic Supply from Wild Stocks

In the United States, the domestic wild-catch of edible products is approximately 3.5 million mt[10], while current U.S. supply of edible products including imports is more than 12.3 million mt. NOAA Fisheries Service statistics[11] reveal that more than 80 percent of our nation's fish stocks are already at sustainable levels (with some yearly variation). Even if all domestic fisheries were simultaneously managed to their long-term potential yield, total supply would be increased by only another 3.1 million mt.

Trend: The Continuation of Policies That Do Not Address Overcapacity Will Plague Both the Domestic and Foreign Commercial Harvesting Sectors

Excess fishing. capacity (fishing capacity is the ability to catch fish or fishing power) and overcapitalization (capitalization, related to capacity, is the amount of capital invested in fishing vessels and gear) reduce the economic efficiency of fisheries and usually are precursors to overfishing. Overcapacity is difficult to manage indirectly, resulting in management regimes that encourage costly and unsafe race-to-fish competitions for limited fishery resources. In 2006, the U.S. fishing capacity of the existing fleet far exceeded the target catch level of many stocks of fish. This overcapacity has reduced economic efficiency and created a race for the fish. In addition, it has negatively impacted the economic livelihoods of many coastal communities dependent on marine fisheries. As harvesting costs continue to rise due to inflation and increasing energy and other business expenses, additional but necessary management restrictions could impact the economic viability of our coastal communities.

The Magnuson-Stevens Fishery Conservation and Management Reauthorization Act (MSRA) in 2006 provided new guidance on the use of Limited Access Privilege programs that directly address the fishery conservation and overcapacity reduction goals of the Nation. Additional new provisions mandating catch limits and catch accountability should improve fish stocks and enhance fishing opportunities.

Trend: The Marine Recreational Fishery Sector Will Continue to Grow as Our Population Grows, Lives Longer, and Has More Leisure Time

Recreational fishing continues to be one of the most popular outdoor sports. Anglers took nearly 93 million saltwater trips in 2005. The increased size of the recreational fishing population creates disputes over allocation of limited resources between commercial fishermen and recreational anglers, and even within different sectors of the recreational community. Technological innovations, however, will continue to assist the survival rate in catch and release fisheries.

Trend: The Contribution of Aquaculture to Supply Fish, Crustaceans, Mollusks and Other Aquatic Resources Will Continue to Grow

Aquaculture will supply an increasing proportion of the world's seafood supply. Globally, aquaculture has increased from 3.9 percent of total fisheries production by weight in 1970 to 27.1 percent in 2000 and 43 percent in 2004.[12] Aquaculture continues to expand more rapidly than all other food-producing sectors. Worldwide, the sector has grown at an average rate of 8.8 percent per year since 1970, compared with only 1.2 percent for capture fisheries and 2.8 percent for terrestrial farmed meat production systems over the same period. Production from aquaculture has greatly outpaced population growth, with per capita supply from aquaculture increasing from 1.54 lb in 1970 to 15.6 lb in 2004, representing an average annual growth rate of 7.1 percent. Today, our domestic aquaculture industry provides 1.5 percent of the US seafood supply[13]. While foreign aquaculture production contributes to an ever increasing proportion of U.S. imports, particularly of shrimp, salmon, tilapia and a variety of bi-valves and mollusks. Total U.S. aquaculture production is about $1 billion annually[14] compared to worldwide aquaculture production of about $70 billion annually. According to the UN Food and Agriculture Organization,[15] global aquaculture production will need to double by the year 2030 to maintain current worldwide per capita consumption. An expanded U.S aquaculture industry can increase the production of fish and shellfish to meet increasing domestic and international demand, assist in fishery stock recovery via enhancement, and decrease the U.S. seafood trade deficit.

Trend: Demands Will Increase for Additional Data and Science Necessary to Support Ecosystem-Based Management.

Humans are components of the ecosystems. they inhabit and use. Their actions on land and in the oceans measurably affect ecosystems, and changes in ecosystems subsequently affect humans. Understanding and modeling this cycle of sustainability of fisheries and ecosystems at an acceptable level of certainty requires a much broader understanding of appropriate and effective science than has been encompassed by traditional, single-species fishery management. Ecosystem research and analyses will increasingly form the basis for new analytical models and assessments of the factors that influence ecosystem status, and predict environmental and social impacts of various management approaches. Using these tools, techniques, and ecosystem indicators, NOAA Fisheries and state and regional management partners will simultaneously be considering multiple objectives, identifying risk factors and uncertainty, and forecasting the cumulative environmental impact of policy choices.

Trend: In the Future, International Fisheries Management Will Have a Greater Impact on the Status of Fisheries Stocks Worldwide

The U.S. government and the U.S. fishing industry are actively involved in the operation of most of the international Regional Fishery Management Organizations (RFMOs). Many of the highly migratory species (HMS) caught by U.S. fishermen in the U.S. EEZ are also harvested in significant amounts by foreign fleets on the high seas. The U.S. government has responsibility to work with other nations to maintain healthy highly migratory and high seas stocks. Eliminating illegal, unregulated and unreported fishing practices is a global agenda. Multilateral policies, standards and guidance on achieving fisheries sustainability will be increasingly common and depend on consumer and market choices, and broader trade and economic sanctions in addition to traditional negotiations to achieve desired fisheries management outcomes.

SECTION 2. MAFAC FINDINGS

Considering the trends discussed above, the following findings and conclusions were reached:

1. Seafood demand will continue to exceed supply even if overfishing is eliminated, current environmental factors which adversely impact stock health and productivity are reversed, and the status of all our domestic wild stocks is optimal.
2. Domestic fisheries alone do not and are unlikely to meet America's demands for seafood.
3. Consumers must have confidence in the safety, quality, and labeling of seafood products worldwide.
4. Limited access privilege programs that protect the fishermen as well as the resource must be established where feasible as quickly as possible with extensive stakeholder input.
5. Recreational anglers will continue to increase in numbers and impact.
6. Sustainable, productive fish stocks and rationalized fisheries will be prerequisites to decrease allocation disputes between and among sectors.
7. The commercial and recreational fishing sectors will continue to play a major role in the economic viability of coastal communities
8. To meet the increasing demand for seafood productsand to reduce our current trade deficit, a robust domestic aquaculture industry must be part of the future of U.S. marine fisheries.
9. Ecosystem-based management approaches will be a major part of the fishery decision-making process.

10. International fisheries will become more important in the future and the United States must continue to be engaged in international RFMOs.

Four recurring themes appeared in almost every analysis and discussion of issues.

1. *Better data are necessary for management decisions.* Every one of the issue areas examined requires more data, more timely data, and data of higher quality to achieve the outcomes desired for fisheries in 2020. In the absence of adequate data, wrong decisions or overly precautious policies will have profound economic and environmental consequences. The current investment in data seems disproportionately low relative to the societal value of the resources under NOAA's stewardship. Fortunately one of NOAA's strengths is in its tremendous capacity for conducting scientific research and collecting data and information. Where this strength turns into value for the public is when the data and science are applied to management policies and decision making.

2. *There are wide-spread opportunities to develop and adopt technology to assist in achieving the outcomes desired for 2020.* Due to the scale and scope of the issues being addressed in fisheries, cost-effective solutions for 2020 will likely involve some form of technology innovation. This will range from: engineering solutions that refine fishing gear selectivity; to improving the efficiency and success of aquaculture production; to adopting low-cost, modular, self-contained sensor packages that can be deployed in various environments to greatly increase sampling range and efficiency for research and monitoring of data required for ecosystem-based assessments. A focused look at internally and externally developed technology's potential role from a perspective other than a single discipline, line office or program point of view could result in substantial programmatic and cost breakthroughs.

3. *Achievement of the Nation's ocean policies in 2020 must result from collaboration and partnerships across levels of government, sectors, and disciplines to advance the ecological, social, and security interests of present and future generations.* NOAA must identify and promote opportunities that bring together different interests and expertise to communicate, coordinate, and collaborate on formulating

sound environmental policies and sustainable ocean management. This will result in the vigorous exchange of science, engineering, technology and policy expertise bothdomestically and internationally.

4. *To obtain these predicted benefits will require additional resources.* Implementing the recommendations for 2020 described in this report will sustain current resource values and, through rebuilding and recovery, will significantly increase the value of our nation's living marine resources. U.S. marine fisheries (commercial, recreational and aquaculture) are an economic engine for the nation.[16] There is a strong positive relationship between the public's interest in proper stewardship of our fisheries and the cost necessary for success. The return on investment for additional funding is high, readily supporting a business case for significantly increasing the nation's investment to satisfy the vision of safe seafood and efficient and sustainable fisheries in 2020.

SECTION 3. SUMMARY RECOMMENDATIONS

MAFAC envisions a future with healthy, sustainable fish populations, a robust fishing and marine offshore aquaculture industry, ample recreational fishing opportunities, numerous, vibrant coastal fishing communities, and a safe and healthy seafood supply for the nation. To achieve this vision, the following recommendations are proposed. (More specific details and rationale for each are found in the Appendices of the report.)

Demand, Supply and Quality of Seafood Products

1. Public health benefits of seafood should continue to be researched, understood and communicated. NOAA should help educate consumers domestically and world-wide about the wide array of health benefits from aquatic foods. The goal is to empower the public to tailor their consumption decisions to individual health needs while reflecting accurate and informed conservation concerns.

2. NOAA should seek both industry and government commitments worldwide to strengthen seafood safety programs, including cooperative efforts through the United Nations/World Health Organization's Codex Alimentarius (food code standards).

3. NOAA should support the federal government's continuation of free trade policies for seafood, and pursue elimination of tariff and non-tariff trade barriers through the World Trade Organization, bilateral and multilateral agreements.

4. Seafood safety and associated human health should be enhanced through improved NOAA enforcement, research, outreach and education, and NOAA should establish itself as an unequivocal source of unbiased peer-reviewed scientific information.

Commercial Fisheries

5. NOAA must achieve and maintain sustainable levels of stocks important to commercial fisheries.

6. NOAA must match fleet capacity with available, sustainable harvest.

7. Limited access privilege programs should be thoroughly analyzed for applicability in all fishery management plans for participants in the commercial and recreational sectors, with the goal of significantly increasing their use by 2020.

8. Commercial fishermen, processing businesses, trade associationsand state and local government representatives working with NOAA Fisheries should seek ways to integrate wild stock production with aquaculture production to maximize the value of domestic seafood production and related industries, including, but not limited to efforts to develop "niche" markets for value added products and wild products. Integration of wild and farmed production can contribute to the development of a stable, year-round processing industry ensuring coastal community sustainability.

9. NOAA must work with states and coastal communities to ensure continued infrastructure necessary to support viable seafood industry along our coasts.

Recreational Fisheries

10. NOAA must achieve and maintain sustainable levels of stocks important to recreational fisheries.

11. Sale of recreationally–caught fish is a form of commerce and should be prohibited at state and federal levels. Improved recreational harvest

data are essential and a recreational registry must be implemented and used.

12. Fishery management plans should include analyses of quota transfer between recreational and commercial sectors, and should incorporate market mechanisms where appropriate.

13. Efforts should be directed to enhance a conservation ethic and pride of a national resource amongst all fishery user groups.

Aquaculture

14. The development of a significant domestic, environmentally sound aquaculture industry is essential for the production of safe and healthy seafood, assisting in the rebuilding of depleted stocks, and providing employment opportunities in coastal communities.

15. National offshore aquaculture legislation providing a coordinated, cohesive and efficient regulatory process should be passed by Congress and implemented immediately.

16. The domestic aquaculture industry should receive financial and technical support similar to that available to the American agricultural industry.

17. Continuous, comprehensive monitoring of offshore aquaculture sites must be included to safeguard wild stocks and assure environmental impacts of facilities are insignificant.

Management

18. Coastal and ocean habitat protection must be a primary concern of fishery managers as a basic requirement for robust and sustainable fish stocks.

19. Ecosystem-based management, including assessments that integrate both habitat protection and multi-species interactions, should be the norm and not the exception for U.S. fisheries management.

20. Cooperative management efforts among states, regional management authorities and federal managers should be maintained and enhanced as a basis for sound domestic fisheries management.

21. Stock status and catch data must be accessible to all stakeholders and provide the information needed to make informed management decisions.

22. Subsistence fishing is recognized as an important source of protein for rural and native communities. However it needs to be included in the calculation of total catch with an efficient, comprehensive, and uniform data collection method.

23. The United States should exert strong leadership in the international forums that manage fish stocks beyond countries' Exclusive Economic Zones.

24. The U.S. government must exert every influence possible aimed at maintaining healthy highly migratory species stocks and barring IUU fish from entering the global market place.

Appendices II-VI of this report contain individual papers prepared by MAFAC members categorized under the following headers: Demand, Supply and Quality of Seafood Products; Commercial Fishing; Recreational Fishing; Aquaculture; and Management. These papers provide more details in support of the conclusions and recommendations noted above. Each paper was prepared using a standard template and reviewed by a MAFAC Vision2020 work group, the Committee as a whole, and circulated for public review.

APPENDIX I. PREPARATION OF REPORT

In September 2006, the Assistant Administrator of NOAA's National Marine Fisheries Service (NMFS) asked MAFAC to prepare a report on the desired future state of U.S. Marine Fisheries. The specific request from the Assistant Administrator to MAFAC was "to create, in clear, simple, non-jargon language, a stakeholders' consensus on the desired future state of domestic and international fisheries."

To meet this request, MAFAC formed a subcommittee composed of MAFAC members to draft a concept paper of what should be included in such a report. The concept paper was circulated and input was received from all MAFAC members regarding a long list of topics to be considered. In December 2006, a MAFAC writing group met in New York to categorize the input received. After review and consideration, the committee organized the input into four subject categories. The committee circulated their proposal to the full committee and the concept and categories were unanimously accepted, and a draft report was subsequently prepared.

The draft report was a major agenda item of the June 2007 MAFAC meeting. By the end of the meeting, MAFAC had reached a consensus on the contents of the report. In August the draft report was transmitted to NMFS, and to receive stakeholder input the report was placed by MAFAC on a dedicated website *Fish2020* for review. At the December 2007 MAFAC meeting all MAFAC members reviewed the public input and collectively revised the report to reflect the accepted comments. This final report reflects the input of all MAFAC members as well as input from various stakeholders.

APPENDIX II - DEMAND, SUPPLY AND QUALITY OF SEAFOOD PRODUCTS

Issue Statement 1: Demand for fish and seafood continues to increase both domestically and worldwide due to population growth, growth of income and growing recognition of the health benefits of a seafood rich diet.

Background: Given the projected population growth worldwide over the next two decades, it is estimated that at least an additional 40 million tons of aquatic food will be required by 2030 to maintain the current per capita consumption[17] Research is expanding our understanding of the health benefits of a diet rich in seafood[18]. If research continues in the same direction, it will likely raise per capita consumption around the world creating an even larger demand for seafood.

*Current Situation:*Americans consumed a record 16.6 pounds of seafood per capita in 2004 and health professionals are encouraging a doubling of the recommended amount to two 6 oz. seafood meals per week. Globally, consumer demand for fish and shellfish continues to climb, especially in affluent, developed countries which in 2004 imported 33 million tons of aquatic food worth over $61 billion.

Preferred State in 2020: Consumers worldwide have adequate supplies of sustainable seafood to satisfy demand for health and nutritional benefits, which are economically affordable and meet personal preferences.

Proposed Actions to Accomplish Preferred State: (a) Educate consumers domestically and worldwide on the health and nutritional benefits of seafood; (b) Continue free trade policies and pursue elimination of non-tariff trade barriers through World Trade Organization, bilateral and multilateral agreements.

Proposed Entity(s) to Promote Action: (a) Department of Commerce's NOAA Fisheries Service and Foreign Commercial Service; (b) U.S. Department of Agriculture's (USDA) Foreign Agriculture Service; (c) Department of Health and Human Services (HHS), National Institutes of Health (NIH) and the Food and Drug Administration (FDA); (d) the private sector; and (e) consumers.

Issue Statement 2: The public is concerned regarding the safety of aquatic foods due to chemical and biological hazards. The public lacks the necessary understanding of the relative risks versus health benefits of a diet rich in seafood.

Background: Seafood causes food borne illness worldwide due to both naturally occurring and handling/processing induced pathogens, toxins and chemical contamination. Seafood safety programs (both public and private) may be inadequate in many countries; yet the U.S. imports over 70 percent of the fish and shellfish consumed domestically. Research over the past 25 years has identified major health benefits of seafood consumption causing health officials to encourage greater consumption (e.g., Americans should double their current seafood consumption levels). However, increases in demand domestically and/or worldwide basis, will place additional stress on seafood safety programs as well as wild capture fisheries.

Current Situation: Seafood safety remains of paramount importance to consumers and public health officials, yet strong seafood safety programs in which the consumer has confidence arelacking.

Preferred State in 2020: Consumers are confident in the safety of both domestic and imported fish and seafood products due to improvements in public and private standards and inspection infrastructure worldwide. Furthermore, more consumers are taking advantage of the health benefits of seafood through increased consumption.

Proposed Actions to Accomplish Preferred State: Effective seafood safety programs, coupled with great consumer education on the health benefits of a diet rich in seafood products, would be a beneficial for health and economic reasons. Both industry and governments worldwide need to strengthen food safety programs, including cooperative efforts through the United Nations/World Health Organization's Codex Alimentarius (food code standard). Consumers are informed about the wide array of health benefits from aquatic foods and empowered to tailor their consumption decisions to individual health needs.

Proposed Entity(s) to Promote Actions: Congress would need to appropriate additional funds at a minimum to strengthen the seafood safety and inspection program. The Administration entities include: (a) NOAA Fisheries; (b) HHS's FDA, NIH and Centers for Disease Control; (c) USDA's Food and Nutrition Service; and (d) the private sector.

APPENDIX III. COMMERCIAL FISHING

The U.S. commercial fishing industry depends upon the long-term sustainability of fishery resources and their ecosystems. Contributing over 35 billion dollars to the Gross National Product, the fishing industry provides an important food source for the nation, creates over 65,000 jobs[19], and affords a traditional way of life for many coastal communities. The U.S. is the world's fifth largest fishing nation and its fleet of approximately 23,000 vessels roams all of the world's oceans. Commercial fishermen nationwide have seen profound changes over time in stock abundance, markets, the stakeholder process, and management of the resource. MAFAC members identified the following four issues to be considered for the future of the commercial fishing community.

Issue Statement 1: Our Nation's fisheries need to be managed to meet sustainable fishery goals, but even if fully achieved they are unable to meet domestic demands for many fish products.

Background: Some marine fisheries continue to be under stress from overexploitation, habitat degradation, or both. Various factors, both natural and human-related, affect the status of fish stocks and their ecosystems. Such factors include: environmental changes, pressure from commercial fishing effort, and loss of habitat.

The long term potential yield of the fisheries within the U. S. EEZ is estimated to be 8.1 million tons per year[20]. However, to reach and harvest sustainably at this level, current efforts to rebuild stocks must be extended to all overfished stocks and rebuilding completed. Efforts to reduce bycatch must be increased. To help meet demand, by-catch and unaccounted mortality will need to be continually reduced to help meet conservation goals. Harvest and landings data need to be improved to account for all mortality. In addition, the current domestic fishing fleet capacity exceeds what is necessary to obtain the target catch level for most fisheries. Fisheries must be rationalized to assure sustainability and protect the fishermen by elimination "the race for the fish." All these measures will be required to approach the long-term potential yield by 2020.

Current Situation: Three principal strategies that are available to or used by fishery managers to manage fishery yields are: regulating fishing effort, restoring habitats, and increasing recruitment. The first two methods are the basis for currently managing our fisheries. Recent landings of U.S. commercial and recreational fisheries are still only slightly more than 60 percent of the long term potential yield. Current management measures are designed to maintain sustainable fisheries stocks, to rebuild depleted stocks to meet the potential long term yield and consumer's demand for fish products.

Preferred state in 2020:

(a) Our Nation's fisheries are actively being rebuilt and are at or approaching sustainable conservation goals.
(b) Technological advancements and market demands have resulted in reductions in undesired bycatch and in increased use of marketable underutilized species.
(c) Our U.S. fisheries are close to achieving long term potential yield.
(d) Coastal commercial infrastructures is maintained or enhanced to support sustainable fisheries and communities.

Proposed Actions:

(a) NOAA Fisheries should consider the role of underutilized species to meet current domestic demand after considering biological, ecological, socioeconomic and technological implications.
(b) Incentives or market development should occur only when research is completed.
(c) Rebuild all depleted stocks by 2020.
(d) Data used for managing marine fisheries must be relevant, reliable, timely, and have stakeholders' confidence.

Issue Statement 2: Some international Regional Fisheries Management Organizations (RFMOs) fail to implement necessary conservation measures to ensure maintenance of healthy stocks, thus reducing the total amount of seafood available to the nation's population.

Background: Many commercial stocks, such as tuna, are highly migratory species which spend most of their life in the open ocean. They are harvested by U.S. commercial and recreational fishermen and by foreign fishing fleets. Although the United States has management authority for several HMS species, most are managed cooperatively by Regional Fisheries Management Organizations (RFMOs).

Current Situation: The performance of RFMOs is uneven, with regard to effective management of stocks under their jurisdiction. This unevenness impacts the U.S. in several ways. First, because the U.S. imports a significant amount of seafood, any mismanagement of stocks on the high seas will ultimately reduce the amount of seafood available for American consumers. Second, because consumers often do not distinguish between poorly managed fisheries overseas and well managed fisheries in the U.S., domestic fishing companies and fishermen can be unfairly accused of inadequate commitment to sustainability. Finally, U.S. fishermen frequently are required to significantly reduce harvests without similar measures being adhered to by foreign fishing fleets. Total harvest reductions are necessary to effectively improve the health of these stocks. The United Nations and the RFMOs themselves are considering means to make the international management of highly migratory fish stocks more effective.

Preferred State in 2020: All fisheries, domestic and international, are effectively managed to sustain long term optimum yields.

Proposed Actions: The U.S. government provides assistance to RFMOs to promote sustainable stocks using available political, economic and other strategic tools to ensure other countries follow the recommendations of RFMO scientific staff.

Issue Statement 3: Overcapitalization has been and continues to be a serious concern in a number of U.S. fisheries. Too many fishermen racing for too few fish has resulted in more restrictive, highly complex and often ineffective management regimes. The race for fish, coupled with other factors has increased operating costs. The result has been lower net economic returns in a number of commercial fisheries.

Background: U.S. commercial landing were relatively stable at about 3 million tons per year from 1935 to 1977 when the U.S. extended its jurisdiction to 200 miles. With the passage of the Fishery Conservation and Management Act in 1976 and other policies, the federal government provided incentives to rehabilitate and expand the domestic fishing fleets. These incentives took two forms: open access management which allowed unrestricted entry to the fisheries, and a number of direct and indirect subsidies to the fishing industry. The goal of these incentives was to ensure full domestic utilization. Since 1977, landings have more than doubled. However, for many fisheries fishing effort grew more rapidly than was sustainable, resulting in overcapacity and in some cases overfishing.

Current Situation: Today, fisheries managers utilize a number of "command and control" management measures to control fishing effort such as limits on fishing days, gear restrictions and trip limits. In addition, most fisheries have some form of limited access. Increasingly managers and fishermen alike are looking at other ways to more effectively reduce and manage fishing capacity including buyback programs, permit stacking programs and limited access privilege programs with assignable fishing privileges.

Preferred state in 2020: By 2020 we will have reached the goal of rebuilding sustainable fish populations while maintaining productivity and biodiversity. This will result in increased biomass, providing greater harvesting and processing opportunities for domestic fisheries and increased supply to consumers. Fishing capacity will be at a level to both efficiently and sustainably harvest domestic fisheries and provide greater economic returns to participants and fishery-dependent communities. Limited access privilege programs (LAPPs) will be in place in most applicable U.S. fisheries, providing market mechanisms to match capacity with available harvest levels.

Proposed Actions:

(a) Commercial fishing interests and other stakeholders should work with regional fishery management councils and NOAA Fisheries to develop regionally-appropriate plans to:
   (1) Reduce overcapitalization; and
   (2) Match fishing capacity to sustainable harvest levels through the use of LAPPs, industry buyback programs and other appropriate mechanisms.
(b) NOAA should play a leadership role by at least tripling the number of fisheries under LAPP management by 2020.

Issue Statement 4: Technology offers a myriad of benefits to fishermen, some of which have significant environmental benefits. In many cases, technology can complement and enhance federal conservation and management goals and objectives.

Background: Many commercial fishermen utilize increasingly sophisticated technology during fishing operations. Electronic equipment common in the wheelhouse today includes state of the art sonar equipment to locate target species, computer logbooks and electronic net sensors. Enhanced sonar capability promotes selective fishing and increases operational efficiencies, including fuel efficiency. Onboard computer logbooks are an important reference tool providing historical catch information and can allow for real-time reporting. Electronic net sensors deployed with the gear can provide important data on proximity to the ocean floor, net profile and the filling rate of fish in the cod end. Each of these technological applications can enhance operational efficiencies and conservation objectives through cleaner fishing and minimizing fishing gear impacts on the environment.

In addition, in recent years many fishery management plans have mandated the use of vessel monitoring systems (VMS) as a management tool.

VMS, or onboard satellite tracking systems, provides managers increased flexibility in developing management measures that can be adequately monitored and enforced.

Current Situation: In recent years, cooperative research involving NOAA Fisheries, the fishing industry, universities and the private sector has produced fishing gear innovations to increase retention of target species, minimize bycatch of non-target species and reduce impact of fishing gear on ocean habitat. The projects are numerous and ongoing, such as: turtle excluder devices (TEDs) in shrimp trawls, chain modifications to reduce flatfish bycatch in the scallop fishery, modified footropes to reduce bottom contact, and technologies to deter seabirds from taking baited fish hooks. Technological innovation is critical in enabling U.S. fishermen to increase efficiency while enhancing selective fishing practices which minimize ocean habitat impacts.

Preferred state in 2020: By 2020 advances in technology wi ll not only result in more sophisticated products, but also the application of the technology can be used for scientific purposes as well as commercial purposes. Advances in gear and monitoring technologies can help obtain information to improve management, reduce bycatch and minimize habitat impacts caused by fishing. NOAA Fisheries is able to increase its efforts to assist in projects that outfit fishing vessels with acoustic equipment that enhances stock assessment capabilities. Also, programs that equip fishing vessels with ocean monitoring equipment is greatly expanded. NOAA's overall science program will be significantly enhanced by utilizing alternative industry research platforms. NOAA Fisheries should continue to place a high priority on expanding its cooperative research program.

Proposed Actions: Actions necessary to achieve the goal of employing state-of-the-art technology in commercial fishing operations to enhance efficiency and promote conservation of living marine resources include: (a) Technology research and development to create more environmentally friendly fishing gear and practices. These designs would improve the performance of fishing gear to help reduce bycatch and minimize habitat impacts, and support additional data collection programs that enhance management, stock assessments and ocean monitoring. (b) NOAA Fisheries and the commercial fishing industry should continue to develop industry partnerships such as its

Cooperative Research Programs and Bycatch Reduction Engineering Programs.

APPENDIX IV. RECREATIONAL FISHING

Issue Statement 1: Growth in populations and coastal tourism are resulting in increasing numbers of recreational fishermen. Therefore, the impact these fishermen are having on fish stocks is increasing. As this demand for recreational fishing continues to increase, recreational fishermen will request increases in fish allocated to the recreational sector.

Background: According to a NOAA report[21], an estimated 153 million people lived in coastal counties in 2003. This population represents an increase of 33 million people or 28 percent from 1980. In addition, a review of NOAA sponsored Marine Recreational Fisheries Statistical Survey data from the years 1981 to 2005 shows a near doubling nationally of marine recreational anglers from 6.9 million to 11.2 million or a growth rate of approximately 3 percent per year. The value of recreational fishing as an economic engine for coastal communities should be recognized and exploited to a greater degree. The recreational fishing experience could rival or exceed recreational fishing catch as a prime motivator for recreational fishing.

Current Situation: The current rate of increase in the angling population creates new management concerns. If the rate of recreational fishermen continues to increase at 3 percent per annum, by 2020 the number of recreational fishermen will increase by 7.3 million to a projected level of 18.5 million. This change will result in a significant increase of fishing effort and catch (i.e., mortality), all else equal. By 2020 continued growth in recreational angling will require that anglers focus more on the fishing experience and less on the number of fish landed. However, while post-release mortality in catch and release fisheries is usually low (often 2-5 percent), as fishing effort increases, post-release mortality will become an increasing proportion of total mortality. It is conceivable that the cumulative total of post-release mortality could increase to levels equal to the total allowable mortality for a fishery. As the number of recreational fishermen continues to increase, improved monitoring will be necessary to assess the fishing effort and catch. A national saltwater angler's registry under development will be a necessary tool to collect data.

Preferred State in 2020: Many recreational species have limited population growth rates and are too valuable to be caught only once. By 2020, catch and release fishing is emphasized and accounted for in specific species assessments. The proper techniques for release are refined and disseminated to lower post release mortality. For other fisheries, minimum size limits and reduced daily bag limits are sufficient management measures to maintain healthy standing stocks. Additional seasonal closures are considered to eliminate or redirect effort. By 2020, angler satisfaction is derived from the recreational fishing experience rather than the take or "kill" fish. To achieve optimum yield, adaptive management measures such as a temporary reallocation of quota is available to managers. For example, if commercial quota is not harvested, managers are able to temporarily reassign the under harvested quota to provide additional recreational opportunity, and vice versa.

Proposed Actions to Accomplish Preferred State:

(a) Improve collection of recreational catch, release, and harvest data,
(b) Create and use the recreational angler registry.
(c) Continue to promote catch and release fisheries,
(d) Reduce daily bag limits and implement minimum or maximum size limits when necessary for those fish stocks where resorting to total catch and release is not necessary,
(e) Promote research to accurately quantify and minimize post release mortality,
(f) Increase the length of seasonal closures when necessary and encourage the recreational community to maximize the profitability of open seasons,
(g) Amend fishery management plans to allow for timely conversion of unused commercial allocation to the recreational sector and vice versa;
(h) Implement a variety of programs and incentives to enhance the conservation ethic of recreational anglers.

Proposed Entity(s) to Promote Actions:

(a) The leadership of the recreational fishing community should promote the total recreational fishing experience, instill a conservation ethic, and de-emphasize landings.

(b) Industry and NOAA Fisheries should continue to support research and technology designed to reduce post release mortality.

(c) Management (councils, commissions, NOAA Fisheries)should consider extending closed seasons to reduce mortality.

(d) Management, (councils, commissions, NOAA Fisheries), should amend fishery management plans to allow, when appropriate, the conversion of commercial quota onto recreational quota and vice versa.

APPENDIX V. AQUACULTURE IN THE UNITED STATES

In 2004, the U.S. Commission on Ocean Policy[22] expressed concern about America's seafood trade deficit and noted the increasing importance of aquaculture products in seafood trade. It noted also that new developments in technology made aquaculture possible in the open waters of much of the U.S. Exclusive Economic Zone (EEZ), where it might now be done on a large enough scale to make a meaningful impact on the trade deficit. Accordingly, it directed NOAA to develop a comprehensive, environmentally sound permitting and regulatory program for marine aquaculture in the EEZ, to which NOAA responded with a 10-year Marine Aquaculture Plan[23] and a proposal for the **National Offshore Aquaculture Act of 2007.**

Issue Statement 1: Growth of American marine and offshore aquaculture should be supported by government and facilitated by providing a coordinated and efficient regulatory system and sufficient funds to achieve this goal.

Background: Development of marine aquaculture in the U.S. is hampered by confusing or overlapping laws, regulations and jurisdictions. Aquaculture operations in offshore waters lack a clear, timely and efficient regulatory regime, and questions about exclusive access have created an environment of uncertainty that is detrimental to investment in this industry[24].

Current Situation: The U.S. has not yet developed the necessary policies for locating, (siting), conducting, and monitoring offshore aquaculture operations. A new governance framework is necessary if offshore aquaculture is to succeed[25]

Aquaculture expansion is supported by the U.S. government, but there is public concern about environmental impacts including possible pollution, escapes, competition with wild fish, disease transmission and food safety. This concern has been heightened by misinformation about aquaculture in the news media[26].

Global supply of seafood from wild-caught stocks has plateaued, while demand continues to increase. Aquaculture now provides 43 percent of the world's seafood. Nutritionists encourage Americans to double their present consumption of seafood to benefit their health.

Preferred State in 2020:

(a) A mature statutory framework will exist for the efficient development of aquaculture in the U.S. EEZ, which protects both the environment and private aquaculture property rights, and provides traceability in the market to protect against the substitution of illegally taken wild stocks.

(b) States will have developed comprehensive nearshore aquaculture plans with technical assistance from NOAA using funds provided by section 309 of the Coastal Zone Management Act. These state plans will protect existing nearshore aquaculture from adverse effects of coastal development and will identify and preserve areas with good potential for future aquaculture development. They will also provide coordinated and efficient regulation.

(c) Aquaculture will be recognized an instrument of national food security policy and will be validated by appropriate incentives and a business climate that encourages good aquaculture practice.

(d) Consumers and the public will be accurately informed about aquaculture and will support sound public policy on its behalf

Proposed Actions: Both statutory and regulatory actions are necessary for a robust domestic marine aquaculture industry by 2020.

Statutory actions:

(a) Develop and codify a statutory framework for marine aquaculture in the U.S. EEZ. Perfect, as needed, the statutory framework for marine offshore aquaculture.

(b) Identify NOAA as the lead federal agency for all offshore marine aquaculture.

(c) Develop economic policies that encourage environmentally sound and economically viable marine aquaculture, include exploring options to promote community and fisherman entry into aquaculture through the use of specific access privileges, cooperatives, and other statutory or regulatory changes

(d) In addition, modify current financial assistance and development programs at the state and federal level to facilitate creation of aquaculture operations similar to the support received by the agriculture industries.

(e) Authorize regional pilot projects involving commercial fishing families to provide a mechanism for fishermen's involvement as well as an educational and outreach tool.

Regulatory actions:

(a) Encourage states to utilize CZMA section 309 funds to accomplish comprehensive planning for aquaculture development in the territorial sea.

(b) Provide sufficient financial support for research and development on all aspects of marine aquaculture including evaluation of best management practices to minimize ecosystem impacts.

(c) Consider establishment of aquaculture zones within the EEZ which would reduce the burden on applicants to submit *new* applications for every proposed project.

(d) Promote outreach and education to enhance public understanding of marine aquaculture.

(e) A Programmatic Environmental Impact Statement (PEIS) for aquaculture projects should consider cataloguing local species and habitat; identifying potential risks to sensitive habitats, fish and wildlife; review of potential wastes, chemicals, and biological pollutants and the anticipated ramifications for local fish and wildlife populations; relevant information on marine ecosystems from the use of feeds; design and placement of aquaculture facilities and expected impact; and expected effect on the human environment including impacts on small businesses and coastal communities.

Proposed Entity(s) to Promote Actions:

(a) Congress for statutory actions with input from the Executive Branch and the public (including industry interests).
(b) State authorities responsible for implementing the Coastal Zone Management Act for coordinating the development of comprehensive aquaculture plans with CZMA 309 funding.
(c) Executive Branch, primarily through NOAA and the Joint Subcommittee on Aquaculture, for regulatory actions with input from the industry, the public, the regional fishery councils, fisheries commissions, and the coastal states.

APPENDIX VI. MANAGEMENT

Based on the current trend, ecosystem-based approaches to management will be the norm and not the exception by 2020. The ecosystem-based management approach is defined as management that is adaptive, geographically specified, takes account of ecosystem knowledge and uncertainties, considers multiple external influences, and strives to balance diverse societal objectives. An ecosystem-based approach to management is incremental and collaborative since the authorities for ecosystem management are distributed across many levels of government, and management requires participation of many different stakeholder groups in public and private sectors.

Ecosystem-based management approaches must be based on high quality, reliable scientific data. For ecosystem-based management to succeed, a significant expansion in the type and quantity of data collected and analyzed must occur. Furthermore, timely accessibility by managers to these new and different kinds of high quality data is critical to success. It is essential to initiate new data collection programs, particularly those utilizing advanced technology, and to expand and improve existing data collection and delivery programs. MAFAC members identified the following issues to be considered when discussing management tools for the future.

Issue Statement 1: Place-based management approaches are gaining acceptance in dealing with a variety of ocean use issues, including protection of unique habitat, location of industrial and scientific research facilities, and conservation and management of living marine resources.

Various state and federal regulatory agencies and private sector interests will become more involved. Traditional fisheries management entities need to recognize the addition of these new and in some cases influential broad based stakeholders.

Background: Marine Managed Areas (MMAs), an example of place-based marine resource management, have been proven an effective tool to supplement traditional management techniques. Examples include seasonal fisheries closures, Marine Protected Areas (MPA's), and No-Transit Zones.

Current Situation: In progress: Number of MPA's and results. Allocations and mitigations/conflicts expected.

Preferred State in 2020: Unique habitats, essential fish or marine mammal critical habitats, and rare marine ecosystems are protected with MMA's developed with stakeholder advice and support.

Proposed Actions:

(a) Place-based management must be better coordinated within NOAA.
(b) If Marine Managed Areas involving living marine resources are designated, they should be based on the best scientific information available.
(c) Criteria for assessing the costs and benefits of closing anarea must be identified, assessed and considered before a decision is made.
(d) The area should be monitored. A timetable should be established for review of the closed area's performance that is consistent with the purposes of the closed area.

Proposed Entity(s) to Promote Actions: (a) NOAA Fisheries should champion place-based management in partnership with NGOs, fishermen and other marine resource stakeholders.

Issue Statement 2: Technology plays a vital role in ecosystem-based marine resource conservation and management and in the development of responsible aquaculture practices. Continued improvements in technology will further enhance sustainable marine resource management efforts.

Backgroundand Current Situation: Technology is integral to NOAA Fisheries' science program, and it plays a significant role in the agency's enforcement and monitoring efforts. Here are some examples of how technology is currently being utilized.

- Satellite imaging assists ocean observation and is an increasingly important tool for assessing fish and marine mammal stocks, identifying "bycatch hotspots," and mapping sensitive habitat.
- In the Alaska region, scientists attach satellite transmitters to marine mammals to collect information on diving patterns. This data is then used to determine the animals' foraging and migratory characteristics, and it assists managers in developing conservation and management measures designed to minimize competition for prey between marine mammals and fishing activities.
- Vessel Monitoring Systems (VMS) employ electronic transmitters on fishing vessels. These transmitters relay information about a vessel's location via satellite. VMS is used not only to enforce management area closures, but is utilized on the west coast for depth-based management for commercial and recreational groundfish fishing.
- Satellite communications assist in fisheries monitoring and enforcement. Federal fishery observers communicate vessel catch data to a central data base on a daily or weekly basis, and this catch accounting is essential to ensure that total allowable catch levels are not exceeded. Also, video monitoring through mounted on-deck cameras is being studied as an alternative to placing observers onboard vessels.
- Work is continuing on state-of-the-art acoustic technology to improve fishery survey work, which is a key component of stock assessment. In fact, NOAA has launched two new research vessels that are among the most technologically advanced research vessels in the world to replace the aging vessels in its fleet, and two more research vessels are under construction.

Preferred State in 2020: NOAA will be utilizing technology to increase dramatically our understanding of the ocean environment, protect and conserve marine resources and provide direct and measurable benefits for the fishing community.

(a) In conjunction with other federal agencies and non-federal partners, NOAA will have implemented an integrated ocean observing system (IOOS), including the placement of biophysical moorings that perform myriad tasks. IOOS systems provide continuous, realtime observations that include acoustic readings that help determine fish and marine mammal migrations and optical technologies that help monitor ecosystem health.

(b) Research in life history, stock structure, brood-stock considerations, spawning, rearing and release of juveniles and ecological concerns will have advanced such that stock enhancement, using hatchery reared juveniles to supplement wild production, is a widespread viable management tool to be considered for rebuilding depleted marine stocks. Research and development of stock enhancement should have expanded such that by 2020 the U.S. can take a role in developing international guidelines and standards. U.S. efforts should have proceeded on a regional basis with a focus on stocks that most greatly impact current and future fisheries management and harvest.

(c) NOAA will be employing Geographic Information System (GIS) tools throughout the country for further improving ecosystem-based management. GIS software allows for visual representation of important ecosystem attributes in map form. Mapping has a number of effective applications for marine resource management, including identifying bycatch hotspots.

(d) NOAA scientists will be routinely utilizing acoustic technology to characterize the seabed. Historically, the process for learning more about seabed composition (a critical aspect of the marine habitat) required removal of core samples. This work technology will also be in place serving NOAA's hydrographic survey mission, working across scientific disciplines to use acoustic technology to perform both habitat research and navigational chart updates.

(e) NOAA Fisheries will be widely employing autonomous underwater vehicles (AUVs), or Seagliders, to enhance its science program. Seagliders are small, free-swimming vehicles that are extremely energy efficient and can be deployed for months at a time. Working jointly with university scientists, NOAA will routinely employ Seagliders to record oceanographic measurements traditionally collected by research vessels, but at much less expense.

Proposed Actions:

(a) Both Congress and NOAA Fisheries should place a priority on applying technological innovations to strengthen science and management programs within the agency.
(b) Future administrations of NOAA should continue the emphasis placed by the current administration on intra-agency and inter-agency coordination of science and technology programs. NOAA's future leadership should also continue to seek partnerships with universities as well as other entities engaged in marine research.
(c) Congress must adequately fund NOAA Fisheries' science and technology programs, recognizing that ecosystem-based management objectives, including an enhanced understanding of the ocean environment, cannot be achieved without investments in technological innovations.

Proposed Entity(s) to Promote Actions: Congress, NOAA leadership, academia.

Issue Statement 3: Allocation disputes currently confound the management of many fisheries. Councils often are faced with making difficult allocation decisions with little scientific information to guide these decisions. Councils should have the option to use assignable fishing rights to resolve allocation issues between commercial and recreational sectors, and within sectors.

Background: Allocation of fisheries between and among sectors has historically been done through political forces exerted on councils or Congress; this has often been a difficult and contentious process. Where assignable fishing rights have been created, market forces rather than regulations have determined fishery entry and exit decisions. Some allocation issues, including those between commercial and recreational fishermen, could be better resolved through limited access privilege programs(LAPs) and all councils should evaluate these mechanisms available to them.

Current State: Individual Transferable Quotas (ITQs) and harvesting cooperatives have enabled industry to consolidate, and provided a mechanism to allocate fisheries to those placing the highest values on the fishery (willing to pay the most). To date these tools have only been deployed in commercial sectors. Acceptance of rights based approaches varies among regions, with strong positions held on both sides.

ITQs are successfully in place on all three coasts of the U.S. Although the North Pacific Fisheries Management Council has successfully implemented ITQs programs for several of its commercial fisheries, the first attempt to implement ITQs for the for-hire halibut sector failed after more than six years of work, due to resistance from the recreational community. Concerns include ability to outbid the commercial sector, ability to pay off their shares, and the potential for migration of recreational shares into the commercial sector. The lack of accurate catch histories complicates initial allocation. Given the proven political clout of the recreational sector, many see it easier and cheaper to compete for allocation through the political process of the councils and Congress, rather than risk allowing market forces to play out.

Preferred State in 2020:

(a) LAPs are widely used in both commercial and recreational sectors to provide the right incentives, address overcapacity and address allocation issues within the sectors and across the sectors.
(b) Reliable catch reporting systems are in place to support stock assessments, fisheries management, and allocation decisions.

Proposed Action:

(a) NMFS needs to work with councils to deploy the new assignable rights authority contained in the MSRA.
(b) Proactive involvement by NMFS with councils during the development stage will help ensure adherence to required processes and standards, resulting in approvable plans.
(c) Continue efforts to promote the value of assignable rights based approaches and publicize success stories in cooperation with the councils.

Proposed Entity(s) to Promote Actions: NMFS, councils, commercial and recreational organizations and other interested stakeholders.

End Notes

[1] The FCMA also created the eight regional fishery management councils.

[2] "Toward Rebuilding America's Marine Fisheries, Annual Report to Congress on the Status of U.S. Marine Fisheries 2006": http://www.nmfs.noaa.gov/sfa/statusoffisheries/SOSmain.htm

[3] "An Ocean Blueprint for the 21st Century", U.S. Commission on Ocean Policy: http://www.oceancommission.gov/documents/full_color_rpt/welcome.html

[4] "Review of the state of world marine fishery resources", FAO report, 2005. ftp://ftp.fao.org/docrep/fao/007/y5852e/y5852e00.pdf

[5] *Ibid.*

[6] "Fish to 2020: Supply and Demand in Changing Global Markets", International Food Policy Research Institute Report, 2003. *http://www.ifpri.org/pubs/books/fish2020/oc44front.pdf*

[7] "Fisheries of the U.S., 2006", NMFS Report, 2007. NMFS Current Fisheries Statistics No. 2006

[8] 2006 Seafood and Health Conference "Seafood is a low-fat source of high quality protein and the health benefits of eating seafood make it one best choices for growing children, active adults and the elderly."

[9] Seafood inspection and assurance of a safe product is becoming a more frequent domestic consumer concern. Congressional hearings and introduction of legislation reflect this growing interest.

[10] "Fisheries of the U.S.,2006". *op. cit.*

[11] "Report on the Status of the U.S. Fisheries for 2006", NMFS annual Report to Congress, http://www.nmfs.noaa.gov/sfa/statusoffisheries/SOSmain.htm#07

[12] "State of world aquaculture, 2006", FAO Report, http://www.fao.org/docrep/009/a0874e/a0874e00.htm

[13] Presentation by NOAA's Dr. Michael Rubino at February 2006 Aquaculture America Meeting: "Offshore Marine Aquaculture: Building on Policy, Technology and Research" *http://www.lib.noaa.gov/docaqua/presentations/aa_offshorepanel_files/rubino_aa_06.pdf*

[14] "NOAA Ten Year Plan for Marine Aquaculture", NOAA Aquaculture Plan: U.S. Department of Commerce, National Oceanic and Atmospheric Administration, October 2007. *http://aquaculture*

[15] "State of world aquaculture, 2006", FAO Report *op. cit.*

[16] With every one pound increase in U.S. fish and shellfish supply, $2.41 in value is added to the U.S. Gross National Product. Non-consumptive and recreational uses contribute billions of dollars to the economy as well.

[17] "State of world aquaculture, 2006", FAO Report: *op. cit.*

[18] See for example web sites of Seafood and Health Alliance http://www.seafoodandhealth.org/ and National Seafood Educators http://www.seafoodeducators.com/home.html

[19] "Fisheries of the U.S.,2006". *op. cit.*

[20] "Our Living Oceans: Report on the Status of U.S. Living Marine Resources, 1999", NOAA Report, http://spo.nwr.noaa.gov/olo99.htm

[21] "Population Trends along the Coastal United States: 1980-2008", 2005 NOAA report, http://marineeconomics.noaa.gov/socioeconomics/assessment/population.html#Download

[22] "An Ocean Blueprint for the 21st Century", *op. cit.*

[23] NOAA Aquaculture Plan *op. cit.*

[24] "An Ocean Blueprint for the 21st Century" *Ibid.*

[25] "Recommendations for an Operational Framework for Offshore Aquaculture in U.S. Federal Waters." Cicin-Sain, B. et al., 2005

[26] "State of world aquaculture, 2006", FAO Report, *op. cit.*

INDEX

Y

W